BOG-STANDARD
BUSINESS

BOG-STANDARD BUSINESS

HOW I TOOK THE PLUNGE AND BECAME THE MILLIONAIRE PLUMBER

CHARLIE MULLINS, OBE

JB
JOHN BLAKE

Published by John Blake Publishing Ltd,
3 Bramber Court, 2 Bramber Road,
London W14 9PB, England

www.johnblakepublishing.co.uk

www.facebook.com/johnblakebooks
twitter.com/jblakebooks

This edition published in 2015

ISBN: 978 1 78418 335 6

All rights reserved. No part of this publication may be reproduced, stored in a retrieval system, or transmitted in any form or by any means, without the prior permission in writing of the publisher, nor be otherwise circulated in any form of binding or cover other than that in which it is published and without a similar condition including this condition being imposed on the subsequent purchaser.

British Library Cataloguing-in-Publication Data:

A catalogue record for this book is available from the British Library.

Design by www.envydesign.co.uk

Printed in Great Britain by CPI Group (UK) Ltd

1 3 5 7 9 10 8 6 4 2

© Text copyright Charlie Mullins 2015

The right of Charlie Mullins to be identified as the author of this work has been asserted by him in accordance with the Copyright, Designs and Patents Act 1988.

Papers used by John Blake Publishing are natural, recyclable products made from wood grown in sustainable forests. The manufacturing processes conform to the environmental regulations of the country of origin.

Every attempt has been made to contact the relevant copyright-holders, but some were unobtainable. We would be grateful if the appropriate people could contact us.

CONTENTS

FOREWORD BY GRAHAM ROBB	*VII*
INTRODUCTION: BLANK SLATE	*1*
CHAPTER ONE: MUM AND DAD	*9*
CHAPTER TWO: GRAFTING	*15*
CHAPTER THREE: BILL	*19*
CHAPTER FOUR: EXILE	*27*
CHAPTER FIVE: SUGAR SNAPS	*31*
CHAPTER SIX: THE REAL APPRENTICE	*37*
CHARLIE ON... WHY YOUR BUSINESS SHOULD TAKE ON AN APPRENTICE	*46*
CHAPTER SEVEN: I COULD HAVE BEEN A CONTENDER	*49*
CHAPTER EIGHT: PLAN B	*57*
CHAPTER NINE: TAKE-OFF	*65*
CHARLIE ON... HAPPY WORKERS	*73*

CHAPTER TEN: INTO THE HOLE	75
CHAPTER ELEVEN: DIGGING IN, AND DIGGING OUT	101
CHAPTER TWELVE: FINE TUNING	91
CHARLIE ON... DOES THE SUIT MAKE THE MAN?	101
CHAPTER THIRTEEN: SAIL STREET	105
CHAPTER FOURTEEN: WHY THE STARS LIKE US	111
CHAPTER FIFTEEN: HOW I GOT FAMOUS	117
CHARLIE ON... EFFECTIVE MARKETING	125
CHAPTER SIXTEEN: BUSTER	129
CHARLIE ON... AGEISM	136
CHAPTER SEVENTEEN: CHARLIE'S ANGELS	139
CHAPTER EIGHTEEN: TV AND ME	143
CHAPTER NINETEEN: THE PUBLICITY MACHINE	153
CHAPTER TWENTY: NOW THAT WE'RE BIG	159
CHAPTER TWENTY-ONE: NOW I'M RICH AND FAMOUS	171
CHAPTER TWENTY-TWO: PLUMBING TO POLITICS	177
THE LETTER FROM THE QUEEN THAT I ALMOST DIDN'T OPEN	189
ACKNOWLEDGEMENTS	193

FOREWORD BY GRAHAM ROBB, CHAIRMAN OF THE INSTITUTE OF DIRECTORS IN THE NORTH OF ENGLAND AND LONG-STANDING FRIEND AND BUSINESS ADVISER TO CHARLIE MULLINS

'Whether you think you can, or think you can't — you're right.'
– Henry Ford, Ford Motor Company Founder

The story of Charlie Mullins is one of the very best examples of a rugged and determined pursuit of business success by somebody from a humble and unprivileged background.

I was introduced to Charlie by a client of my firm who had supplied his depot with high-speed rolling doors allowing his vans to enter and exit quickly without the loss of heat to the depot. The company concerned, Union Industries based in Leeds, was used to supplying warehouses with these doors. The specifications concerning the speed and safety of the doors are always exact. My client, though, was taken aback by the layout, cleanness, and precision of Charlie's depot.

I was told that I really must go to see Charlie and that I would be amazed by his operation – which has since grown at least fourfold. Indeed I was completely gobsmacked! Something so apparently conventional as a depot for vans at a plumbing company was a shining example of military precision, order, efficiency and calm. If plumbing depots were hotels this was a seven-star luxury establishment in Dubai.

As a business owner myself I found it inspiring. No van was allowed out on the street without being perfectly clean. Any little ding or dent was hammered out and re-sprayed in the attached body shop. All tools were tidied and put in their place at the end of a shift. The floor was painted and you could eat your chips off it. This was a company whose dedication to service started at the very bottom and progressed to the very top.

Moving to the then fledgling call centre, desks were tidy, call sheets neatly completed and dockets properly filed. The entire operation was just smooth. Phones were answered in a timely manner by pleasant, well-informed operatives. Customers were given accurate and carefully considered times that their jobs could be undertaken.

The plumber's canteen was pristine, mugs of perfectly blended tea served at clean and tidy tables to plumbers whose 'uniforms' were checked every morning and wore shirts and ties and proper shoes.

Was this a plumbing business or the lair of a Bond villain with obsessive-compulsive disorder who wanted to take over the world with an army of henchmen in blue vans? It was easy to be impressed and easy to pause and think it peculiar. But the end result is magnificent. If you are a nice family in London who has a burst pipe, you will make one phone call and in good time a smartly dressed, well-trained plumber will arrive at your home. He or she will not cause further damage, bring more dirt

FOREWORD

into your property or suffer from a lack or equipment, skills or support. They will be well mannered and efficient. In a world in which we routinely expect problems with service, Pimlico Plumbers has instilled quality into every level of its operation.

In the 1980s this used to be known as TQM – Total Quality Management. It then became something that was assessed with standards such as Investors in People and ISO9001. However, at Pimlico Plumbers independent assessments are not required. The standard is set by Charlie Mullins. It is a high standard and comes from within the personality of the man at the top.

After meeting Charlie I was excited about representing him and wanted to take some of the things I learned back to my own business (some of the changes I implemented are novel in the world of PR – prices that are displayed on my website, for example!)

Charlie is a positive thinker; he personifies the attitude of Henry Ford. He has taken something that is not at all unique, plumbing and home maintenance, and was one of the first business leaders in this sector to apply common systems and clear thinking. The white van man plumber is still there but so is the Savile Row dressed businessman who understands the need for innovation, efficiency and customer care.

The story you are about to read is a tale that might not have come about. Charlie freely admits that without the guidance of people he met in business he could have easily drifted into a life of crime. He will also acknowledge the role of community organisations – such as his boxing club – in shaping his life.

Despite being brought up in a home with no money Charlie did have the right value structure and an instinct for helping others. I have seen the kindness he has exhibited towards so many organisations and charities. As somebody who has worked for and received the rewards of success in business, he has not hesitated to help when he has seen a way in which his money, skills or ideas can make a difference.

The causes and organisations he helps have something in common; they are about people here in the UK who can be supported quickly and effectively. The Prince's Trust, charities helping people with disabilities, funds for sick children, boxing clubs that provide a place for young people to train and gain a sense of the rules of life. The money and time all make a difference, as the stories detailed later in the book will demonstrate.

On a personal level I can tell you that Charlie really does have a heart of gold. I live in the North of England and in 2014 my wife, who coped with cancer for many years, took suddenly very ill and within ten days had died. When she came home from hospital for her final days her room was filled with flowers provided by Charlie. As two fellow business owners the relationship between us became one of friendship and fellowship at my time of grieving.

Charlie is also an instinctive free-market Conservative. Like me, he was a great admirer of Margaret Thatcher and, as a result, he is outspoken in his politics. His was the first London business to openly back Boris Johnson for Mayor of London. As the economy improved after the great recession he argued hard for the funding of proper apprenticeships. He took his message to David Cameron and George Osborne, who both got to know and respect Charlie's opinions.

Very few business people in this country have the courage to

FOREWORD

speak out about politics but Charlie's views are instinctive and invariably credible and correct. In private I know he is sometimes a bit nervous about remembering key facts but does possess the courage of his convictions in daring to speak out despite the potential it has for attracting criticism. It was audacious to say the least that he went on *Question Time* and battled against seasoned politicians to put his point of view across.

This book is two stories for the price of one – a business story and a personal story. For anybody, like me, who has started and grown a business, the experiences Charlie had at the beginning of Pimlico Plumbers are both familiar and heroic. The fear of the first office rental, coming to terms with the need not just to do your job but to work on the business, the exhilaration of business growth and the feeling that you really are 'King of the world!' Then a crash down to earth as external events hit you; in Charlie's case the recession of the early 1990s and the familiar and well-trodden path followed by banks to limit their exposure by killing off their own customers. In this part of his story Charlie proves he is a fighter and a survivor and ultimately a winner. Do not start a business before you read these chapters!

Charlie's personal story is also a tale of beating the odds; he came from a poor family living in a tenement flat, messed about at school and at the age of fifteen could well have taken a direction that set him up to fail. But instead he gained a proper indentured apprenticeship, four years with a City & Guilds qualification. He describes this with the wisdom of passing years and, quite rightly, wants the Government to turn back the clock so that these apprenticeships become more widely available again. They gave young people a sense of personal discipline in addition to solid and practical skills, which set them up for years of higher earnings.

The apprenticeship Charlie had also involved a mentor, Bill Ellis. Somebody whose advice and candour gave Charlie the strong direction and inspiration he needed to tackle the decision points of life. Mentorship is something Charlie has a strong belief in. His support for the Prince's Trust involved him in mentoring young people in addition to the plumbers and staff who, as part of an extended family in business, have been given the benefit of Charlie's wisdom over the years.

On a personal basis I have cause to thank Charlie for his support. My own firm has been built over a similar timescale and has faced the same type of issues as Charlie. As a PR firm it has different challenges and being based outside of London – in the North East – I often find my London-based clients a bit sniffy about our location. Charlie, though, takes a different view. He knows that the London economy is the engine of economic growth in the UK and, like Mayor Boris, he is clear that London's resources can power the entire country. Just as the engines for London's new red buses are made two hundred yards from my office, Charlie's core-PR is prepared in Darlington (in conjunction with our colleagues at PHA in London). This willingness to see the larger impact of his business on the wider world is something that gives Charlie a sense of his relevance and economic importance. His story shows that he understands how the world is linked together, how politicians might dream their dreams but they still need cold cash to deliver them. That the country can only succeed if it has successful businesses

FOREWORD

to provide the cash and, most importantly, that pennies from heaven have to be earned here on earth.

This book, told in Charlie's own authentic voice without the need to polish the prose, is a must-read story of our times.

INTRODUCTION

BLANK SLATE

When David Cameron first invited me to Downing Street, he said, 'Charlie, what should we do about youth unemployment?' And I said, 'You really want to know?' And he said, 'Yeah, of course I do.' So I said, 'It's simple.'

First, the Jobcentres: blow them up, every single one of them.

Next, stop giving kids Jobseeker's Allowance (JSA). Instead, give the money to companies so they can take on young people as apprentices. And I don't mean take them on as slave labourers or for useless work-experience placements. I mean proper, old-fashioned apprenticeships, so that when they're done they can do something useful and earn a living.

And David smiled and said, 'Yeah, interesting.' There were other people there, too. Other 'bigwigs' like me! It was quite a fancy reception.

That was in 2011, after the Coalition got in, and the Jobcentres are still there and loads of money is still getting poured down the drain on Jobseeker's Allowance, and the number of young people unemployed and dropping out of school and college is still far too high.

It's a shambles.

So I don't think David really heard me. Or maybe he heard me but in his mind it's more complicated than that. Me, I'm a fairly straightforward kind of person — 'black-and-white', some might say. But I just don't think it's all that complicated. That's because I started out in life poor, avoided school like the plague, and could have been heading straight for a useless life. But I was lucky enough to get a place as an apprentice plumber when I was fifteen, and it made me the man I am today.

I was born in a horrible tenement house in 1952. This was Camden Town, London NW1, long before it was desirable. My mother and father, three brothers and I shared one bedroom. My parents were alcoholics; they didn't care if we went to school or not. At least I had two parents, a roof over my head, and food, which is more than a lot of people in this world can claim, but you could say I was born into a no-hope part of society.

When I was eleven we moved up in the world. We finally got a council flat at the Elephant & Castle, on the sort of estate they should burn to the ground, the sort where as a kid if you wanted to get along, you had better join a gang. School was a joke; I hated it.

Can you picture the sort of life I was headed for?

Well, now I'm a millionaire. I'm a millionaire plumber. I went from being a North London street urchin to owning the

BLANK SLATE

city's biggest independent plumbing firm, which I built myself from scratch. We turn over more than £22 million a year, and we're growing. The country's best plumbers and heating engineers queue up to work for me. We are the plumbers of choice for London's high society – movie stars and celebrities of all kinds. I'm mates with David Cameron, George Osborne and Boris Johnson.

How did that happen?

I was always a grafter, but when I was nine a local plumber gave me a job helping him out. His name was Bill Ellis. I was going to say 'after school', but the fact is I'd often bunk off school to help him. I owe a lot to Bill. After we moved I lost track of him but I owe so much to him that when I started putting this book together and thinking about it all I thought about hiring a private investigator to track him down, so I could thank him.

Here's why. When you're nine you're a blank slate, unless you're totally screwed up. You are completely open to receiving messages about the world and how you should live. Bill took me on and taught me things. But the plumbing was not the most important thing: the most important was self-respect, the value of hard work, and this blinding bit of advice:

'Charlie,' he said, 'if you learn a trade you will make lots of money and you will never be out of work.'

A very basic thing to say, I know, but I was nine, so it went in deep and stayed there. Even now, when I take home over a million a year, I still say it to myself. Economic turmoil, natural disaster, bad decisions and worse luck could all take my wealth away in a flash, but as long as I can carry tools I'll be the best plumber in London.

BOG-STANDARD BUSINESS

I didn't get organised until I was fifteen, when I left school. My biggest regret was that I didn't leave at fourteen. I dimly remember them trying to get me to learn logarithms, whatever they are. They also tried to drum Latin into me. I'm aware that the word 'plumber' comes from the Latin word *plumbum*, for lead, as in lead pipes, which I suppose might fill up thirty seconds in a flagging pub conversation. Apart from that, by fourteen, I could read. I could also add, subtract, multiply and divide, which are the only things you really need when it comes to handling a pound.

At fifteen, salvation arrived. I was offered a place as an apprentice plumber with a building firm.

These days apprenticeships are a joke. The word 'apprentice' has been used by companies who flaunt it for PR purposes to inflate work experience placements, job-shadowing, shelf-stacking and other forms of cheap labour. It has also been bent out of shape by Alan Sugar's stupid TV show, in which celebrity-crazed wannabe 'entrepreneurs' try and ingratiate themselves to a Lord by scratching each other's eyes out in a televised cage fight.

Each false representation does damage to a time-honoured institution that faces extinction, an institution that is far less glamorous but infinitely more valuable to individuals and society than the bastardised 'apprenticeships' dished up by the media and government propaganda today.

I believe the word 'apprenticeship' should be protected the way 'Champagne' and 'Camembert' are protected, and must mean what it used to mean, and did mean for me: a rigorous, mutual commitment between a young person and a company, where the end product is a skilled, work-ready young adult.

When the riots broke out in the summer of 2011 I felt dread.

BLANK SLATE

Thousands of kids were looting, starting fires and running amok in the city. Would I have been amongst them forty years earlier, in the summer of 1971? Well, I can tell you without a doubt: I would not. And I remember it well. I could taste the end of my four-year apprenticeship, I was hungry to get out on my own and earn some proper dosh. I had a girlfriend and we planned to get married. I wanted a better flat, I wanted a holiday, I wanted a nice life and I had a plan for how to get it. There was too much at stake for me to go acting the hooligan all hours. I could get arrested; I could get hurt. How would I explain it to the older guys I worked with, who depended on me? On top of that, I was too bloody tired – my apprenticeship was a full-time job, and I was in training to be a pro boxer.

Now picture a different Charlie Mullins, aged nineteen in the summer of 2011. Still in the old bedroom, dropped out of school. Last time he checked, if he checked at all, McDonald's wasn't hiring. Parents down the pub. He's staring at *The Apprentice* on TV, wondering what this fantasy world is all about. Phone bleeps. It's his mates, saying: get out here, it's a riot! Literally! What's he going to do?

Right now almost a million kids in Britain between the ages of sixteen and twenty-four are 'NEETS', meaning Not In Education, Employment or Training.[1] Saying 'NEETS' feels stupid so I call them cast-offs, because that's what they are. Schools don't know what to do with them, colleges don't know what to do with them, and companies don't know either.

---

1 Office for National Statistics, 21 November 2013: For July to September 2013 there were 1.07 million young people (aged from sixteen to twenty-four) in the UK who were Not in Education, Employment or Training (NEET).

BOG-STANDARD BUSINESS

You might be thinking, okay, a million. Is that a lot? Yes, of course it's a lot. It's a million kids waking up every day, wondering what they're going to do. A million kids growing up useless.

Is it more than other countries? Yes, a lot more. The only countries worse than us at looking after their young people are Greece, which is an economic basket case, and Spain, which isn't as bad as Greece, but a lot worse off than us.

Britain doesn't know what to do with its young. We shouldn't be up there with Greece and Spain in the cast-off-your-youth league table. Other countries understand. Take Germany. If you're a teenager in Germany you're not waking up every single day wondering what to do with yourself: you're in school or out working as an apprentice, learning something valuable and making money. Germany and other countries, like Switzerland and the Netherlands, decided they were going to do something about this, and set up systems. We chose to ignore it.

You may be thinking, okay, so what?

Well, it doesn't take logarithms and Latin to grasp what's going to happen if you let a million kids grow up useless. And a million after that. The riots were just the start. Britain's welfare bill will keep going up, so too will its prisons' bill, and its policing bill, and its courts' bill. There's going to be this huge section of society that are cast off and don't give a stuff. And it's growing, this cast-off section. Youth unemployment has been rising faster in Britain than in most other countries you'd want to compare us to.[2]

That's why I'm writing this book. Partly it's because I want

---

[2] 'International Lessons: Youth Unemployment in the global context', The Work Foundation, January 2013, page 10.

BLANK SLATE

to describe how it all happened — my business, my success — because it wasn't all that complicated or lucky, and I hope people find it useful, in case they want to do something similar themselves.

The bigger reason is that I'm tired of looking at the shambles. We're crap at looking after our young and we have to get a grip. Apprenticeships are the way to go. The apprenticeship system we have now is stupid: it's cobbled together from this and that and choked with red tape. It does offer money to companies but I can't be bothered with it. Meanwhile, I have twenty apprentices right now learning and working at my company, Pimlico Plumbers, in everything from plumbing and engineering to motor mechanics and business administration. Some of them stay on and work for me, others don't. Each one costs me £15,000 a year. I do this because it helps me find good people, and because it's the right thing to do.

I talk to a lot of business people, big-business people and small-business people, and the weird thing is, everybody I talk to has job vacancies and big problems finding the right people to fill them. They're always moaning about how young people don't have any experience or qualifications. It's like there are two ships, one filled with British youngsters and the other filled with British employers, and they're sailing off in opposite directions.

The system is broken and to me it's quite straightforward how to fix it.

So, David Cameron, this is for you.

CHAPTER ONE

MUM AND DAD

I wasn't looking forward to this bit.

There's nothing frightening or dramatic to report. It was all just tedious, depressing and grim, which is what the lives of the poor are like.

As I said, I was born in a tenement house in 1952, in Camden Town. Mention Camden Town today and people think, yeah, cool place. Lots of bars and restaurants and hipsters hanging out at Camden Market. Famous people lived there, didn't they? Amy Winehouse? Noel Gallagher? Pulp? Loads of bands played at the Roundhouse back in the day, right? What a great place to be from! Well, I can tell you that no, Camden Town was not a great place to be in the 1950s. It started out as a gentrified sort of place in the country after the Regent's Canal was built in the early 1800s, then it got cut off from everything by the railways and became an industrial zone, which went into decline when roads and motorways made the canal and railways obsolete. During the Blitz it was bombed to pieces and thousands of people died – I read somewhere that one stretch of bombing lasted fifty-seven days in a row. I still remember the bomb sites, big craters where houses or shops used to be. All that death and destruction was still a very recent memory when I was born. A lot of struggling families lost husbands and brothers in the war.

Camden was crowded and poor, and of course you had all the stuff you get in places like that. Drinking – there was a pub on every corner, four or five in between and one or two up every backstreet. There was prostitution, thieving, gambling, and gangs. By the 1950s all the big houses originally meant for middle-class families in the 1800s had been turned into tenements, chopped into the tiniest spaces and rented out to people like us. We had one room for cooking and eating, and another for sleeping – my mum and dad, three brothers and me. No bathroom, no hot water.

As I've already mentioned Mum and Dad were alcoholics. It seemed normal. Mum worked as a cleaner and a barmaid. She would do a shift cleaning, then pull pints so she could spend afternoons and evenings in the pub.

Dad had a job at a place where the workers painted little toy cars. After work it was down the pub. He was in the Army during the Second World War, fighting in the Middle East and maybe other places. There weren't the words for him or the bond between us for me to pick up much besides that.

And so they grafted and plodded and drank. Dad was around more. As he was taught and expected to do, he'd pull out his belt if we caused havoc in the house. After Mum died of Alzheimer's

MUM AND DAD

at sixty years of age — they say it was brought on and made worse by the drinking — he hinted to me that he drank because she drank, that he'd never see her if he didn't. In other words, if you can't beat 'em, join 'em.

I never fell out with my dad, but there was a distance between us. You wouldn't say anything to him if you'd done something wrong because the belt would come out. And you'd do the same if something went well because in his mind nothing *ever* went well. I'm rich now and even kind of famous yet we've never even spoken about it.

Committed working-class drinkers who are unskilled don't have a lot of spare cash. And remember, this was post-war Britain, the original 'Age of Austerity'. We went around in plimsolls stuffed with cardboard to keep the rain out. One of my jobs was to queue up outside the second-hand shop on a Monday when the new stuff came in. Then my mum would come and look for clothes. There wasn't much to choose from in those days. One day I even went to school dressed as a German soldier!

Committed working-class drinkers don't have a lot else to offer, either, like role models, guidance, self-respect, introductions to other sorts of people, or the idea that maybe things could be different for us, that they could be better for us, if we put our minds to it. Mum and Dad didn't bother with us going to school, so we bunked off whenever we felt like it. School was where you went when there was nothing better to do.

When I was eleven we moved to a council flat on the Rockingham Estate, at the Elephant & Castle. Dad kept his old job, but his day got a lot longer because now he had to get up to North London by Tube and train. Mum picked up the same sort of work — there was a pub on every corner in those days.

In one way this was an improvement. We had hot running water, an indoor toilet and a couple of bedrooms. But it was horrible, too. We didn't call them 'gangs' back then, but there were people you could knock about with and people you couldn't, so you had to stick to your own territory. There always seemed to be family rows and people fighting outside. I don't know if there was more crime than in Camden, but people seemed to get along better where I'd come from — it was just a very unhappy environment. Years later I went back there to shoot part of a television programme and I noticed there were bars on the windows. I remember seriously wondering if those bars were there to keep people out, or to keep them in.

I would never bring friends round, because of the drinking and also because it was a neglected house, with nothing nice in it. I hated it, and was always having rows with my mum and dad. When I was sixteen I left home, moving into a council flat my nan had, because she was living with her daughter most of the time. This was my mum's mum. She'd been moved down to Elephant & Castle, to the same estate we were on, just after we came. And she hated it too, so she went back up to Camden as soon as she could. We weren't close or anything, I think she just wanted someone in there to keep an eye on things. But she was canny, or tried to be, and that was something we shared. I remember being over there one day and someone knocked on the door. It was the television licence people. I was in the kitchen and I heard her let the guy in.

'Have you got a television, ma'am?' I heard him say.

'Well, yes, I do, but it don't work,' she said. 'It never has.'

'Would you mind just trying to turn it on for me?' he continued. 'Just so I can tell them back at the office that I've checked.'

'Why of course, dear,' said Nan. 'Hang on.'

My mind raced. The last thing she needed was a fine. Then I spotted the fuse box on the wall next to the boiler. Without thinking, I jumped up and switched off the mains. Then I ambled into the lounge, casual as you like, where Nan was bending over and fiddling with the television. She seemed confused.

'See? Just like I told you,' she said, straightening up. 'It don't work, and it never has.'

'Well, thank you very much, ma'am, we'll be sure to make a note of that. And if you do get it repaired, don't forget to get a licence.'

And off he went.

'That's peculiar,' Nan said to me when he'd gone. 'Normally what I do when they come is switch the telly on but turn the sound off, so it looks like it don't work. But now it really don't work – he must have put a curse on it or something.'

'No, Nan, I just switched the mains off,' I explained.

'What, just now? In there?' she said, and then she grinned. 'How clever!'

So that was Nan. Anyway, nowadays, if one of my grandkids said they were moving out at sixteen I'd have a heart attack, but I thought nothing of it at the time.

I said I wasn't going to sit here and bad-mouth them, but I'm not sure I succeeded in that. After all I had two parents, a roof

BOG-STANDARD BUSINESS

over my head, and food, which is more than a lot of people in this world ever get. Both Mum and Dad had genuine work ethics, in their way. In those days there wasn't a lot of support for people like them. They did what they did with the cards they were dealt, and the four of us brothers tumbled out like dice, all very different, some rolling high, some low.

Let's talk about something else now.

CHAPTER TWO

GRAFTING

School was where you went when there was nothing better to do, and I had plenty of better things to do, because I was grafting.

It started when I was little. After school I used to go to the local shops and ask if they needed any errands doing. I'd go into the greengrocer's and say, you know, do you need anything? And the lady there would say, yeah, go and get me some cigarettes. In those days there was only ever one person behind the counter and they couldn't get out. So I'd go and do that and get a thruppenny bit off her. And when I was in the sweetshop, getting the fags, I'd say to him, do you need anything? And he'd say, yeah, get me some milk. So I'd go to the dairy and while I was there I'd say, what do you need? And he'd say, can you get me some bananas, and on it went like that. And they'd be saying, aw, thanks very much! Meanwhile, I was cashing in. I thought

BOG-STANDARD BUSINESS

I was so clever with it. They don't know how I'm doubling up here, I thought.

I also used to do a milk round for the dairy on a Saturday. At five in the morning I had to trundle the bottles round in a barrow and climb up loads of stairs in these old mansion flats in the dark. I was terrified because there were no lights. I'd carry the bottles right up to the top, five, six, seven storeys, then fly down, grabbing the empties on the way, desperate to get out of each block.

My other Saturday job was delivering 'bagwash'. People would give their laundry to this guy who took it to a shop to be washed and dried, and then he'd deliver it back to the families, up in these mansion blocks. I was supposed to be just helping out, although it was always me running up to the top of the stairs. We were always in a hurry because he'd double parked or something, and he used to say, 'Go on, let's see how fast you can get up there', and off I'd run. He'd never do it. I used to think, 'You crafty so-and-so.' But the thing was, when you delivered the bagwash the woman of the home would give you a tip. He never paid me, there were just the tips.

On weekends also I used to go around with mates carrying buckets and knock on people's doors, asking if they wanted their cars washed. That was always successful. You could count on some cash, though you had to get into parts of the city where people had cars – we were the boys' bucket brigade on the Tube!

A couple of nights a week I also worked in a fish-and-chip shop. They had a machine for peeling the potatoes, and my job was to get the eyes out, which the machine couldn't get. I used

GRAFTING

to work there with a mate of mine from school. We thought it was great, but we used to keep getting electric shocks. I'm not sure why, but I think it was because there was so much water around, coming off all these potatoes getting peeled in the machine, and there were cables snaking around on the floor. We'd wear wellies and everything but we'd still get shocks. We loved it! Can you imagine any of that happening today?

This was the part of my life I loved. I had money in my pocket, which went on sweets and knick-knacks. Once I went into a department store and came out with a boy's wristwatch with a real leather strap! It cost five shillings — a lot of money for a boy like me at that time. I was shaking when I walked out of the shop.

The money was great but the real attraction was getting out of the house, talking to people, showing off, having a laugh — just being involved. Nowadays boys can't do this. Even if they could be torn away from their TVs and Xboxes, getting kids to work for you like that would land you in court. It's a terrible shame. Society is scared of child labour, sending kids up chimneys and all that. Fair enough, but the result is that they are cut off from the world and the people around them. They're locked into this artificial universe that is some busybody's notion of what a good childhood is. I don't think it equips them for how the world really works.

Anyway, up to this point I was just doing this and that for cash, which made me happy and kept me sane, but I was about to walk through another door into a whole new world, and this would be the making of me.

17

CHAPTER THREE

BILL

It's funny how random people can rise up out of the general background and start playing a central role in your life.

Bill Ellis was known around the area as a good plumber, a man about town, with plenty of cash, driving a motorcycle, and someone you wouldn't want to mess with. He wasn't a bully or anything, and he wouldn't take liberties, nor would he let insults or nonsense go unchallenged.

Before I started working with him he had a reputation because one time some youngsters, local tough guys, were taking the piss out of him because he wore glasses. He went up and had a go at a couple of them. It was verbal only, but he was quick-witted and he made them look stupid. That night he went to the local pie shop, where everybody used to hang out, and they were there. He said, 'Look, we had a bit of a falling-out but I don't want to be against you, boys. I'm a nice guy, let's be friends.' It was an offer they couldn't refuse.

BOG-STANDARD BUSINESS

There was a lot of snobbishness around in those days, and tradesmen were often looked down on. One time when I was working with him we were in this posh shop and he kept getting ignored by the guy behind the counter.

'Watch this,' he told me.

To the shopkeeper he said, 'I think I'm next.'

'I'm serving,' the guy said, right snotty like.

'Okay,' said Bill, 'but I am in the queue.'

Finally he got served but when the guy gave Bill his change, Bill gripped his hand hard and mashed his fingers. I didn't know what was happening but I could see the guy's eyes widen with alarm and pain and Bill just kept smiling at him. 'It's nice to be treated with such respect in a quality shop like this,' he told him, and let go. We walked out and the guy never said a word. I mention these things because Bill had self-respect and knew how to stand up for himself in all kinds of situations.

We got to know each other, I guess, because we were both blokes working around the place. He'd have seen me delivering laundry, or running errands or in the chip shop, and if we passed on the street we'd have a chat and a joke. He was friendly and easy with people. One day he asked if I'd like to help him out with a job after school the next day. I remember being so excited I could hardly sit still. The job was to fix a leaking water pipe in somebody's back garden. I helped him dig and expose the pipe. It was magic, how he worked out exactly where the pipe was by reckoning from the stopcock, and then where the leak was by feeling the ground. We dug, and there was the pipe, and there was the leak.

'Just tap it,' he said, handing me a hammer.

BILL

I tapped it and the leak stopped. Magic!

'That's it?' I asked.

He laughed. 'No, that just stops it for a bit. Now we fix it.'

After that I helped him three or four days a week, sometimes after school and sometimes I'd bunk off school. The deal was he'd pay me half a crown a day, plus my dinner, which was always egg, chips and two slices of bread and tea.

As I said before, when you're nine you're a blank slate, unless you're totally screwed up. You are completely open to receiving messages about the world and how you should live.

Bill took me on and taught me things. For one thing, I knew how to change a ballcock by the time I was ten. He'd say right, take that thing out and put this thing in. And it would work. And the woman would be, you know, 'Oh, that's amazing! This toilet hasn't worked properly for years!' and I'd be thinking, 'Wow, I did that!'

Sometimes we had to put a whole new toilet in, and he'd always give me the job of making the last connection so that, in my mind, it was me who'd done it. He knew the bits that would influence a child and he wanted to build me up. 'Put that up there, right, now turn the stopcock,' and it would start working. And I'd have this big smile on my face, and so would he.

But the plumbing was not the only thing he taught me, nor was it the most important thing.

I also learned how to get inside the heads of customers.

The name of the game was getting a tip, and working fast was the way to get one. He knew customers didn't want to know all the ins and outs of how their plumbing worked; they wanted

BOG-STANDARD BUSINESS

the problem fixed quickly and they didn't want to get ripped off. So we'd come out, brushing off our hands and say, there you go, all fixed. And the guy or the lady would say, 'What, already? How extraordinary!' And more often than not they'd give us a tip. Sometimes it would be half a crown, sometimes a whole pound (seven or eight days' wages for me). That didn't happen often but when it did, it was like we'd won the pools. Bill would always give me a share.

He also taught me street smarts. It was remarkable how he could predict what people would do. When we dug up a pipe or did some other mucky work my clothes would get dirty, and at the time I didn't have too many spare pairs of trousers. 'Don't worry about that,' he said the first time. 'When we're done the lady will look at you and say, oh dear, what will your mother say? And you say, you ain't got a mother.'

And that's what the lady did say, and when I said I didn't have a mother, she gave me a big tip.

I'm not saying he was a saint. Sometimes if we were working on the outside toilet he'd say he had to nip off for a bit.

'What if they come out?' I said the first time.

'They won't,' he insisted. 'It's too cold. You just stay here and every once in a while give this a bang and tap on that a few times.'

He went out through the garden gate. I don't know what he was doing, getting materials, having a crafty cuppa, seeing some girl, whatever. I was so nervous it must have sounded like a steel band from the house, but he was right, they never came out.

He looked after me, too. There were nonces about and they

were a lot bolder than they are these days. Once he was putting something together in his van and he told me to go and get us some sausage rolls from the shop. On the way back this guy started talking to me. Did I like sweets? Would I like to go to the zoo with him one day? I did like sweets, and the zoo sounded great, but I said I had to check with my boss. 'Okay,' he said, 'you let me know.'

Back at the van, Bill asked who I was talking to. When I told him he said, 'You go back and tell him that if he doesn't give you five pounds right now, I'm going to knock him down.'

So I went back and told him, and the guy ran off before I'd even finished delivering the message.

It was a while before I figured out what all that was about.

Bill taught me many things, including how wrong you could be in your thinking about something. For instance, he had cash and he was liberal with it. If he came across a down-and-out he'd give him money; he'd buy anyone a drink or a meal. He also showed off with it. Never to those who had no money themselves, but as a tool to bring people down a peg or two. If we were in some fancy cafe and they were snooty about tradesmen he'd find some excuse to get his wad out. He'd peel off a note, give it to me and say, 'Remember to get those fittings,' or something.

Anyway, I was convinced he was a burglar. There was no way he could get all that money from plumbing. The plumbing was a front, I thought. He'd leave bathroom windows open and come back at night and rob. I pictured his customers in posh neighbourhoods, Primrose Hill, Hampstead Heath, coming downstairs next morning after Bill had been to find their houses empty – everything gone, even the furniture.

BOG-STANDARD BUSINESS

So in the beginning I used to watch him. If he asked me to open the bathroom window I'd think, 'Aha, that's it! He wants me to do it and leave my fingerprints all over the place.'

It took a while for me to realise how stupid this idea was: he was successful because he had lots of repeat business and word-of-mouth recommendations. He could never have the reputation he had if he was doing people over, no matter how clever a scammer he may have been.

I worked with Bill for two years. My mind was so open that anything he taught me went straight in and stayed there. Of all the things he taught me, the most important was this bit of advice, which never left me:

'Charlie,' he would say, 'if you learn a trade you will make lots of money and you will never be out of work.'

From day one he drummed that into me. Three pounds was what he charged for that leaking pipe in the garden, my first job with him. Afterwards he showed me the money, a wad of ten-bob notes.

'Do you know how much your father earns in a week?' he asked.

I did know. It was three pounds. I didn't say anything but he could see my eyes go big.

'You remember that,' he told me.

I know now that when he said 'money' it wasn't exactly just about the money. Yes, he meant the money. Money is great — money is always great — but the deeper thing, the thing that money is only a symbol of, is self-respect, belonging, standing in the community, being able to look after yourself

BILL

and others, and being able to contribute. That's why what he said was so powerful.

It was a great two years, but then it all came to an end.

CHAPTER FOUR

EXILE

I hope it's clear now why the move south to the Elephant & Castle was a complete disaster. Yes, we had moved up in the world – more bedrooms, our own bathroom – but I had to give up my jobs, and my jobs were all that made sense to me. They were all that was good about my life. It was like going into exile. My world was blown apart.

I was only eleven!

My secondary school career got off to a terrible start. Mum and Dad left the whole school thing up to me, including finding the school, and I managed it as any school-averse eleven-year-old would – I didn't go. I had a look at a few, but guess what? I didn't like them so I spent a whole year hanging about. But I did find a job, eventually: I used to set up tables for a market in the mornings and take them down in the evenings. And I started going back to Camden to do the milk round on the weekends.

Eventually somebody kicked up a fuss – from the council, I think – and I had to go to school. So I ended up in this boys' school. I never really liked school anyway but this was terrifying compared to the friendly, mixed primary school I occasionally went to, back in Camden Town.

I just didn't see the point of school. Now I look back and wonder what was I learning? I remember messing about with logarithms. Did people use these in the real world, I wondered. Where? How? Bill never mentioned them. We never did logarithms on the milk round or in the chip shop either. History lessons? While I think it's good to know a bit of history, it goes on and on. Learning it is a twenty-four-hour job.

I liked metalwork and I liked woodwork. They were subjects I could see the point of – I loved working with my hands. But there was hardly any chance to do those things because most of the time you had to sit in some classroom listening to all this unnecessary stuff.

A couple of the teachers I liked. One or two had something about them – mostly, I suppose, because they took a bit of an interest in me. Otherwise the school didn't know what to do with me – I was just a square peg in a round hole. That's why I felt stupid, though nobody was there to tell me then. Now that I'm an adult I can say it and not feel ashamed. I don't believe in square pegs in round holes, it's a waste of time.

And I have to be honest here and say that I didn't know what to do with school, either. Kids get along in school for different reasons. Having friends really helps, and there weren't many people I could call friends. I was new to the place and it was pretty rough. And I didn't have a happy home life either so I didn't feel all that safe or welcome anywhere. I read somewhere that children of alcoholics tend to bury their feelings. I'm no psychologist, but that was probably true for me. Maybe it still is. Anyway, whether it was me or them, I didn't mix in. The fact that I spent a year not going to school says a lot – I simply preferred to be on my own.

Now I'm a patron of The Prince's Trust and we take a lot of young people in for work experience placements. My heart goes out to them because I know how they feel: they think they're stupid. Some of them have had a terrible time, terrible home lives; some need discipline and straightening out. Others are just square pegs, and schools don't know what to do with them.

Maybe it would have been different if my parents understood what schooling was all about, took an interest in it, and encouraged me. Then again, maybe not. Maybe I just would have been a happier bad student. Better adjusted. Complacent.

To be honest, the question doesn't interest me. Here's why: by the time a boy or girl reaches fifteen there's not much you can do to fix their upbringing, but there are things you can do to help them take their place in the world. That's what I'm interested in.

I want to say one more thing about Bill, because (thank God!) we're about to leave my childhood years. Basically, I lost track of him, which is a sore spot on my conscience. I always thought about him, and still do, but I never made the effort to look in on him, never a phone call, a Christmas card, nothing.

Why?

I tell myself that it's because my life got busy and hectic, and I'm just not good at keeping in touch, and all that. But the reality is, I think I know what he would have said to me.

He would have said: 'Aw, that's great, Charlie! You've done real well for yourself. I'm very proud of you. Now, tell me, have you looked after your mum and dad?'

And the answer would have been, no.

What he would have meant was, Charlie, did you stay involved in your parents' life? And did you make sure, with all your wealth, that they were happy? That they shared in your good fortune with nice things and special comforts in their declining years?

And the answer would have to be, no, Bill, I did not do that.

I don't picture Bill asking why not – he would be too polite to do that. But I can picture the look of disappointment in his eyes. Him kind of closing down.

So why didn't I?

Because they were doing okay, my parents, I suppose, as things went. And they were distant from me, and I didn't want to be involved with them. Because I didn't even know how to go about being involved with them. And life always seemed too complicated and hectic for that sort of thing, and when there was money it didn't seem to be enough to go spending it on other people.

There are probably more excuses, Bill, but what it comes down to is this: my heart turned out to be not so big as yours, on that front.

But I'm learning.

CHAPTER FIVE

SUGAR SNAPS

Salvation came when I was fifteen. It was 1967, and I was taken on as an apprentice plumber. This was a real, proper apprenticeship, and it saved my life. But before I tell you about that, I want to tell you about the very public ruck I had with Alan Sugar, back in August 2009. I want to tell you about it because it will help set out where I'm coming from on apprenticeships.

Now, I respect Alan Sugar as a family man, as a businessman, and as someone who does a lot for charities, but his television show is awful. I don't just mean it's ridiculous, and in bad taste. I mean it sends the wrong message out to young people about the workplace. It isn't a frightening, intimidating place with someone hollering at you all the time. It takes three years to train an apprentice, while on the show you have a bunch of wannabe entrepreneurs running around, trying to please a lord with stupid get-rich-quick schemes and if they hit on the magic combination in a few episodes they get a great whacking salary.

Most of them I wouldn't trust with a broom at Pimlico Plumbers.

I feel very strongly about this because when the young people from The Prince's Trust come to have a look round Pimlico Plumbers, do you know what their eyes are filled with? No, it's not wonder, or suspicion, or gratitude: it's fear. Bald terror. For at least 90 per cent of them it's the very first time in their lives they have ever set foot inside a place of work, and they are petrified.

Now I'm not blaming Alan Sugar for that look in their eyes, not all of it. But the kind of show he's fronting doesn't help – it only increases the false mystique and remoteness of success. One person with some special talent gets all the glory and the rest are losers. For young, poor people television is the main source of information about the world. How does it go? 'Twelve tough weeks, one life-changing opportunity…' The show plays into the same mentality that says, why should I get some useless job and slog my guts out when thieving and drug dealing could make me rich and powerful?

People say, relax, Charlie, it's just the name of a TV show, and I say, no, the word still means something important – something ancient, in fact. I wouldn't mind if the series was called *The Bum Licker* or *Silly Corporate Wannabes* or *Half-wits Who Think They Can Get Rich Quick* – anything but *The Apprentice*.

As I said before, I believe the word 'apprentice' should be trademarked and protected in the same way as Champagne and Camembert, and must mean what it used to mean, and did mean for me: a rigorous mutual commitment between a young person and a company, where the end product is a skilled, work-ready young adult. With a million cast-off young people in Britain today – a number that ain't going to get any smaller on its own – this is a deadly serious issue. You could call it my line in the sand.

That line was crossed in 2009. In June Gordon Brown made Alan Sugar the Labour government's 'Enterprise Tsar', whatever that was supposed to mean. He became the public face of Labour's useless apprenticeship scheme. In August that year the media were saying the British Government had spent £2.85 million trying to fill around 18,000 vacancies. And the result of that nearly three million quid? Around 5,000 enquiries, or so they claimed, and a pitiful 1,185 actual apprentices in place.

So I kicked off. I put out a statement pointing out that if one of his half-wit 'apprentices' had blown three million quid getting 1,000 bums on seats the old finger would be up, pointing, and he'd bark, 'You're fired!'

It was beautiful! The papers picked it up. It gave me a chance to shout about my very simple idea, which is that the government should stop paying Jobseeker's Allowance to under-twenty-fives and give that money instead to companies to pay for apprentices. The money would pay the full salary of apprentices for the first year, to get them started. In the second year, the company and the government could split the salary fifty-fifty. And in the final year, the company would pay the full salary because by then the apprentice would be pulling his or her weight a bit.

We worked it out that with three million quid we could have helped thousands more into apprenticeships and paid for the wages of hundreds of these for the next three years.

It was a tragic waste.

I said some other things, too. It got a bit personal. Anyway, it touched a nerve because the next thing I knew, I got the letter of doom from Sugar's lawyers, Herbert Smith as they were then, the big-shot international firm, saying he's going to sue me if I don't apologise. Sugar snapped. I wanted to fight it out so I went to my then publicist, Max Clifford, and asked him what I should do. He said, 'Charlie, back down. Do what he wants.'

So I apologised, and gave some money to a charity, which Sugar also wanted me to do. And that was that.

Meanwhile, apprenticeships in this country are still a shameful joke.

I remember Gordon Brown making a fuss about apprenticeships in 2008. This was after his buddy Tony Blair came up with the nutty idea that half of all Britain's kids should go to university to study things like golf course management and surfing. Suddenly, apprenticeships were on the agenda. So Labour tackled it the way they tackle everything – blowing trumpets, setting targets, establishing 'task forces', and making other useless gestures such as hiring Alan Sugar as 'Enterprise Tsar'.

And the result has been shambles. Almost anything now is an 'apprenticeship'. Kids can do a six-month work placement and it gets called an 'apprenticeship'. Politicians crow about the surge in numbers of apprenticeships but it's just more window dressing. Seventy per cent of 'apprentices' now are not real apprentices, but existing employees. And most of these new 'apprentices' – 94 per cent, I read – are over the age of twenty-five according to 'The Husbands Review of Vocational

Education and Training', Labour.org.uk, page 2. Companies, and government departments are just reclassifying junior workers for PR purposes and to get some of the cash available from the government.

The problem is, British business started getting out of doing real apprenticeships from around the 1970s and now there just aren't enough places to soak up the young people. Companies don't want to know. According to the National Apprenticeship Service, more than 1.4 million online applicants competed for 129,000 vacancies posted in the twelve months to 31 July, 2013, and that's up 32 per cent on the previous year.

It is now easier to get into Oxford University than it is to get an apprenticeship in some cases. In 2012, a kid seeking a place at Oxford had to compete with an average of five others, while in the same year Rolls-Royce reported 4,000 applicants for just 200 apprenticeship places – that's twenty applicants per place.

The apprenticeship system we have now is stupid. Anything can be called an 'apprenticeship' these days, from two-week work experience stints to shelf-stacking slave labour. There are so many loopholes in the system many so-called 'apprentices' are just existing employees being upskilled. In 2012, the BBC's *Panorama* reported that one in ten 'apprenticeships' created in England the previous year was at supermarket chain Morrisons, who used public funds to put existing staff through six-month training programmes. Most of them were over twenty-five. There are all sorts of other problems with the system. Where do I start? How about this: the funding rules penalise the over-nineteens. Apprenticeships usually involve on-the-job experience and time in college, learning the theory. Up to eighteen years of age, apprentices get all their college costs covered by the government, but that drops by half on their nineteenth birthday, so the employer has to pay. That would be fine if the nineteen-year-old was productive by then, and starting to earn his keep. But these days, kids mature a lot later than they did when I was training. That's not me being a grumpy old bastard, I honestly believe it's true. For many youngsters today, especially the ones who don't do well in an academic setting, the world of work is totally foreign. They just need more time. Often they don't really start pulling their weight until they're twenty or twenty-one.

Many companies can't afford this extra cost. Take the building trades, which have been the mainstay of the British apprenticeship system for centuries. Up to 60 per cent of all apprentices in construction are employed by tiny firms with less than ten employees, according to a cross-party parliamentarians' inquiry report called 'No more lost generations', published in February 2014. These companies carry the burden of training for the whole industry, and they're the least able to afford it.

The system is also a bureaucratic mess. It's cobbled together from all sorts and choked with red tape. We tried it: they wanted a bunch of their people to sit down with a bunch of our people. There would have been a bit of money for that, with hoops to jump through, and in the end it would have taken up so much of our HR and finance and people's time that we couldn't be bothered. So we said 'No, thank you' and just continued doing it our own way, stumping up for it ourselves.

What this country desperately needs is a return to real apprenticeships. I will now tell you what I mean by that.

CHAPTER SIX

THE REAL APPRENTICE

So, salvation came when I was fifteen. I was lucky enough to be offered a place as an apprentice plumber with a company called Anglo Scottish Construction, over at Raynes Park in south-west London. I say I was lucky. That's because even then, in 1967, good apprenticeships were not so easy to come by. The long decline had started and there weren't that many on offer.

By then the school had pretty much given up on me. The careers teacher was a miserable bastard. He was like, so, what are you going to do, Mullins? Everybody else seemed to have it sorted. Some were going on to college. Others had found jobs, working in shops and the like. The school had helped them with introductions and interviews, but I was an awkward sod; I wasn't going to work in any shop. I kept talking about plumbing, and I guess the careers teacher didn't know any plumbers himself or how it all worked, and couldn't be bothered to find out. Even back then the construction trades were seen as second class. People used to say, 'Oh, you'll end up on a building site', meaning nowhere. Which kind of sums up how bad the system was, and still is, at getting young people into work.

Then one day, out of the blue, the careers teacher came up to me and said, 'Mullins, you're a very, very lucky boy.' He had found a company willing to take me on as an apprentice plumber. It was all very sudden. The next day I had to get over to Raynes Park for the interview, and I had to bring my mother. We met the managing director. This guy sat behind a desk, getting strangled by his tie. I'm not sure whether or not I had ever been in an actual office before, but it felt very strange and intimidating. It was also weird, having my mum there. Up to then I'd managed my own affairs pretty much solo, for better or for worse, but she had to be there because we would be signing a legally-binding contract and I was a minor.

I don't remember much about the interview, but he offered me the job. He said the rate was a shilling, 11 pence and a ha'penny an hour, and then he made a song and dance about rounding it up by the ha'penny to 2 shillings. I remember thinking at the time, 'You mean, smarmy bastard, making a big deal about that! Do you think I'm stupid?'

Never mind. My working life had begun, and not a moment too soon.

I'd always been a grafter but it was a shock adjusting to this new regime. Five days a week I had to get up at dawn to catch a bus to Waterloo, then ride the train to Raynes Park, and then walk for twenty minutes at the other end to get to the Anglo-

THE REAL APPRENTICE

Scottish offices. It was a two-hour journey each way. There I'd join whatever plumber I'd been assigned to for that day's work. (Fortunately, the company gave me a travel allowance. Otherwise, with all the bus and train fares, there wouldn't have been much left over from the £3 a week they paid me!)

It had been four years since I'd worked with Bill and I was out of practice, but that didn't stop me being cocky at first. After all, I'd been able to change a ballcock at the age of ten! It wasn't long, however, before I was taken down a peg or three by the old hands. Bill had been generous and easy-going but here I had to earn the respect of the plumbers and grow into the role I'd been offered.

Mostly, for the first year-and-a-half that meant twisting into awkward places to work rusty bolts and fittings loose, doing all the stinky, horrible jobs, humping sacks of rubble down endless flights of stairs, and lugging heavy tools, pipes, grouting and whatnot back up.

'Just nip down and...' became the four words I hated most.

Once, I complained. Under my breath I muttered, 'Why do I always have to do this?' or something.

The plumber stopped what he was doing and looked at me, hard. 'There's the door,' he said. 'You know how to use it.'

Cheap labour, that's what it felt like, and one part of my fifteen-year-old self rebelled. I was right, too. To get their money's worth, Anglo Scottish needed an unskilled dogsbody like me to help the plumber so he could get on with the skilled bit. But there was another side of it that I didn't appreciate, which is that to learn a trade you have to learn all of it, and you have to start from the ground up.

BOG-STANDARD BUSINESS

Being a good plumber is one thing, but making money at it is another. Successful plumbers don't just offer a skill, they offer a service. By watching and mucking in and being a slave I learned how to approach a job, how to keep a tidy worksite, how to organise tools and materials, how not to waste time and money, how to behave, how to handle customers, and a hundred other things you can only learn by watching and doing.

Plumbing, I would eventually come to understand, is only a bit about plumbing.

It got so that I loved the journey home. Every time my bum hit a seat I was asleep. Regimentation, focus, discipline – it was all exactly what I needed. I pretty much disappeared from the estate. Weekdays I worked, evenings I boxed (I'll tell you more about that later), and weekends I caught up on sleep. People – by that I mean loafing mates with nothing else to do – used to say, 'Charlie, I don't know how you do it!' But to me it was worth it. All the while I could hear Bill as clear as a bell: learn a trade, earn lots of money, never be out of work.

And despite my punky attitude with the MD, for that first year I was proud of the money I was bringing home – nearly £3 a week. That was a fair amount of cash for a fifteen-year-old who already had a roof over his head. But in the second year I started to resent it, especially after I'd moved into my nan's flat, which made me think I was all grown up. Three pounds a week? It started to feel pathetic. That's because by then mates of mine were making more. By the time I was seventeen, eighteen, some guys I knew were earning twice that working as painters and decorators, or in shops. One worked in the John Lewis department store on Oxford Street.

THE REAL APPRENTICE

He was in the white goods department, selling dishwashers and washing machines and all that. With commission thrown in, he was sometimes bringing home £12 a week. Four times what I made!

I thought, 'Charlie, you mug, you've got it all wrong! You've made a mistake.' It used to eat me up – these flash gits, waving their paycheques around. But the older guys I worked with would keep saying to me: 'Charlie, just stick with it. You watch, their money ain't going to change, but yours will.' It was hard to take. I was still a foolish kid in many ways. When you're that age you want everything and you want it now; you can't see the bigger picture.

And of course my workmates were right. Those guys working for fly-by-night painting and decorating outfits, labouring or doing other low-skilled jobs, they hit their pay ceilings pretty much in the first couple of years. Soon after I qualified as a plumber I was making twice, three times what they were, and I was working for myself. But it was tough on the pride. I got married to my sweetheart Lynda when I was nineteen. (We'd known each other since I was fourteen or fifteen. We didn't go to the same school but we used to knock about with the same people in those days.) This was before I'd finished the apprenticeship. Lynda worked as a telex operator and when I showed her my paycheque the first time I was embarrassed because she earned a lot more than I did!

The thing was, though, I couldn't just up sticks and quit. That was the beauty of it. I'd signed a contract and if I'd broken it, there'd have been hell to pay. I couldn't have finished my City & Guilds course – there'd be a black mark against my name. Other companies wouldn't have hired me and I'd never be able to set up properly on my own. The system would have spat me out.

But it went both ways: the company had obligations, too. If I came in late or didn't show up for a course one day they couldn't just fire me. I'd get a bollocking — and you didn't want too many of those — but I'd still have my job. It gave me a bit of elbow room for growing up, to learn the trade but also learn how to be an adult.

They say it takes a village to raise a boy, and the apprenticeship system was the nearest thing you could get to a village. Yes, you had to do all the rubbish jobs to help the tradesmen — fetch tools, bring tea, tidy up, and whatnot — but they had to reserve some of their time for bringing you on, giving you tasks, looking at what you were doing, and showing you the right way if you did it wrong.

There's another saying. It goes, 'Tell me three times and I may remember, show me once and I'll never forget'. A skilled craft like plumbing takes a bit of theory, a bit of book learning, but you learn so much more when you do it and get instant feedback. With practice you develop the knack for things. People like me who get cast-off from the academic education system, we're not stupid — it's just that we learn with our eyes and our hands. We learn from real-world cause and effect, from the look and feel of things.

Then the classroom stuff makes sense. The college bit of it I loved. There I was, in a classroom, surrounded by tools and machinery! We learned how they worked, what they did, and how to take care of them. And we were taught by geezers who

THE REAL APPRENTICE

knew their stuff; they let us into the secret world of materials. Things we couldn't do before – building things, fixing things – we could now. Suddenly, school had a point. It was appealing. We'd go to college for six weeks at a time for classroom and practical training. This was to give you the next bit of knowledge you needed so that your employer had a little bit more to work with in making you useful. I loved it, but it took some getting used to. The first exams I failed. We had practical exams and theory exams. I sailed through the practical but failed the theory. There were six or seven of us who failed, and lo and behold, the failures were amongst the six or seven who spent most of their time messing around! I guess a lifetime of bunking off school was catching up with me. After that I knuckled down and applied myself.

I said at the beginning that the old apprenticeship system was a rigorous, mutual commitment between a young person and a company, where the end product was a skilled, work-ready young adult. The men understood that – so did Bill Ellis. The time it took and the obligation to be patient was wired into the system. This much I know for sure: without that framework I'd have quit a thousand times, and been sacked a thousand times, too.

I'll give you an example. Two years into it I had a row with this old plumber they'd put me with. Old-school, very proper and set in his ways, he got on my nerves if I'm honest. One day we were threading some pipes. You had to knock the die out with special little hammers. It was a delicate operation, something you develop a knack for, and I got impatient. I started whacking the thing like the stupid kid I was, and chipped a bit of metal off. It was an expensive and time-wasting mistake. He went off on one and we had a row and I walked out. Then I went back to the company and said I didn't want to work with the guy any more.

I was in the wrong and he was a very good plumber, but they decided to give a little leeway to a hot-headed teenager and so they put me with a younger plumber. I'll call him Eric. He was great at first. The others thought he was a bad influence, and he probably was. We'd do things like bunk off and go to his house for a cooked lunch. But in other ways it was a valuable experience. They had us working on a building site and we both hated it, getting shouted at and bossed around, fighting for space with the other trades – building sites work on a divide-and-rule basis. The whole thing was horrible. I decided I would never work on building sites, and to this day Pimlico won't touch the construction side of things.

Meanwhile, Eric taught me every trick there is for skiving. We'd spend all day avoiding the foreman. Once we were installing these big tanks and we climbed into one, pulled a piece of ply over the top, lit a Bunsen burner to keep warm, and spent hours chatting and thinking how clever we were.

Eventually the novelty of Eric wore off. I'm not a skiver, and it was starting to do my head in. After a few weeks with Eric I went to the City & Guilds people and told them I didn't want to work for that company any more. I guess I didn't have the confidence or the diplomatic skills to work it out with the boss directly. They scratched their heads for a bit and then came up with a brilliant plan: it wasn't the done thing, but they arranged for me to go and work for another company that was much closer to home, F&H Plumbers in Loughborough Junction, just down the road from the Elephant & Castle.

Overnight, life got a lot easier. I could sleep in! Not only that, F&H taught me a whole different side of plumbing. These days I'd like to see it written into apprenticeships that you have to work for two companies, at least. They said things like, haven't they taught you how to wire a pump? Haven't they taught you this? Haven't they taught you that? And I thought, 'This is great! I'm getting two complete lots of training for the price of one.'

For instance, in a lot of ways F&H were a much slicker operation. They had more business savvy and later I would apply parts of their approach to Pimlico, such as insisting on payment as soon as the job was done instead of invoicing the customer and walking away, which adds a whole layer of hassle in chasing payment and invariably causes cash-flow problems... but I'm getting ahead of myself.

When I turned nineteen I was ready to leave, and they were ready to let me go. There was an unwritten rule at the time that apprentices should be allowed to go when their time was up, if they wanted to. Otherwise the danger was they might always be just 'the boy' around the place. Now I'm not saying it would have been like that at Anglo Scottish, or at F&H, they were both good lots of people. But by then it was clear to me, and probably to them, that I wasn't cut out for taking too many orders: I had to be my own boss.

At fifteen I had left school with no qualifications, and my biggest regret was that I didn't leave at fourteen. I look at my apprenticeship as one of the most important things that happened to me in my life. It's up there with marrying Lynda, having my kids, working with Bill, doing *The Secret Millionaire*, and one more thing, which I want to talk about now.

Because, the thing is, I have another confession to make.

For this to be like the movies I should have been completely single-minded in my ambition to be a plumber. I ought to have been humming the *Rocky* theme or something as I lugged pipes up the stairs. But the truth is, by nineteen what I really wanted to be was not a plumber at all.

> **CHARLIE ON... WHY YOUR BUSINESS SHOULD TAKE ON AN APPRENTICE**
>
> Apprenticeships hold the key to solving two of the economy's major ills, an increasingly ingrained youth unemployment problem and a growing skills gap that will leave us trailing behind international rivals in terms of competitiveness and productivity.
>
> There's no denying apprenticeships are fashionable again, although, mainly amongst politicians and young people. Of course, that's not a bad thing. Young people have realised that there is more to life than academic study and the government knows that vocational training delivers actual job-related skills that can drive the economy forward.
>
> The challenge is creating enough opportunities for apprentices to flourish from trainees into valuable members of the workforce. The only way this can happen is for more employers to offer apprenticeship places.
>
> Some firms, of course, place apprenticeships at the heart of their skills strategy and even sectors that wouldn't previously have been associated

with apprenticeships, such as law and finance, are catching on to the benefits of training people 'on the job'.

However, plumbers' maths will tell you it's just not enough.

A Demos/British Gas report revealed there are just eleven apprentices for every 1,000 employees in England compared to thirty-nine in Australia, forty in Germany and forty-three in Switzerland. Now don't get me wrong, this isn't some stick-beating exercise blaming employers for the skills gap and lack of opportunities for young people. We've all been through the toughest economic storm for several generations and times have been hard. As a result investment in the workforce has dropped off, which includes apprenticeships.

This has created a skills gap, which means we are playing catch up. Fewer than 10 per cent of firms in England offer apprenticeships and young people are unable to find placements due to the number of applications from the growing army of school leavers keen on the vocational route. So, businesses have to look long and hard at their future and, even if they can't see immediate returns from employing apprentices, they must be able to appreciate what'll happen to their businesses five or ten years down the line when the flow of skilled workers dries up.

Of course, businesses can't do it all on their own and with all the will in the world the money just isn't there to create the thousands of apprenticeships needed. The government is trying to improve the situation and the current Skills Minister Matthew Hancock and his team deserve praise for their efforts to improve the standing and availability of apprenticeships, but more needs to be done. This is why I am continuing to call for the creation of a new training allowance using the money currently being given to out-of-work young people in the form of Job Seeker's Allowance. Divert that money to employers to help fund apprenticeships and I am in no doubt that the number of apprentices would skyrocket.

It's so simple and provides a win-win situation for employers, young people and government. Creating more places would have a direct and positive impact on the economy. Apparently, an additional 300,000 placed would boost the UK's GDP by £4 billion as well as cutting the youth unemployment rate. Those are numbers not to be sniffed at.

So, this Apprenticeship Week, if you're an employer who can afford to take on even just one more apprentice, do it, you won't regret it. If you can't, join me in my calls for the creation of a national apprenticeship training allowance. Either way it's a no-brainer.

Originally published on 3 March 2014 on realbusiness.co.uk — the UK's leading title for high-growth businesses and entrepreneurial SMEs.

CHAPTER SEVEN

I COULD HAVE BEEN A CONTENDER

What I really wanted to be was a boxer. From the age of fifteen to twenty-one, when I was forced, unfairly, to retire, boxing was absolutely the number one priority in my life. In my mind I was going to be a pro, and I was going to be famous, and if my chance hadn't been cruelly taken away from me I'd have dropped the plumbing in a shot. I probably would have dropped the wife and kids, too, I was so committed. I'm not proud of that but it's how it felt at the time.

I came to it through a guy at school, Tony, who got bullied all the time. He had a sweet nature and was timid, but I liked him. One day he was being set upon by this bunch and I flew into a couple of them. Off they went. Tony was all, you know, 'Thanks, Charlie, and all that', and then he said to me, 'Charlie, do you like boxing?'

'What do you mean, boxing?' I said.

'You know, boxing,' he said. 'Down at the gym. I do it, you should come along.'

I thought he was taking the piss, but I went along to have a look and blow me if he wasn't the best boxer in the club! He was hot! Up there in the ring nobody could touch him.

After, I said, 'Jesus, Tone, why don't you, you know…?'

It made him happy to hear that but he just sort of shrugged and looked sheepish. Maybe he didn't want to let on too much in case it made him even more of a target. Or perhaps he was afraid of what he could do. Maybe he could only turn it on when he was in the ring. Whatever, it was a mystery to me.

Anyway, seeing him really inspired me, so I joined the gym, at the Fisher boxing club near London Bridge station, and started training.

I took to it like a duck to water. Coming from a poor background, the Rockingham Estate and all that, and having two older brothers, I was no stranger to fighting, but boxing lets you channel all that instinct. It gave form to it, science and style. I was like, 'Wow, this is the future!'

People who know me know I'm no dabbler. It's all or nothing, and boxing was more 'all' than anything before or, probably, since. Six, sometimes seven nights a week, I'd be at the gym, training. Most mornings I'd run. That was two sessions a day, which is too much because you can burn out, but I didn't care.

At first I was crap, then I was okay, and then I was very good.

My first proper fight, I lost, and it gave me an early warning on the politics of boxing. I was sixteen, matched against this guy who was twenty-eight. It was his last show appearance and he was from a local club. We went the rounds and he wasn't all that good but the judges gave it to him. I'd have had to knock him out in round two to win that fight and even then he'd have won on some technicality.

It left a sour taste. I realised some fights you've lost before you even get there. But it made me train harder and be more aggressive. I was damned if I was going to let the judges make the decision for me again! We had a pro boxer help out sometimes at the amateur club and I used to spar with him. Something he said to me ramped up the killer instinct immediately.

It wasn't nice.

'Charlie,' he said, 'when you go into the ring you should be thinking, "I want to f***ing knock this geezer's head clean off his shoulders".'

Then he said: 'And when it comes off, you're going to kick it out of the room like a football.'

It wasn't nice but I ate it up. It's the mindset you need if you want to win, and I wanted to win. I wasn't there to prance about, having a laugh. It's terrible to say but I would have killed someone in the ring if it meant winning. Nowadays I'd rather lose than leave someone injured but then – and I mean playing by the rules and all – if someone died I'd be like, who cares? I won!

My next thirteen fights were either wins or knockouts. When you have a string like that it gets attention, and I started appearing in the local papers and boxing magazines. It was my first taste of fame, if you like, and I took to it like a duck to water, too. I remember one of my older brothers saying he'd read about it in the paper, and he said, 'Hey, Charlie, seems you're good at this, you should make a go of it.' Of course I was already doing that, but hearing it from him was a big deal for me.

Before too long I was representing London in tournaments. I represented London against Germany. That was an eye-opener. I think the farthest I'd ever been at that point was Luton, and there I was, in Düsseldorf! And we were treated like kings. We stayed in a nice hotel and we were taken to see places. The Germans were intimidating – they took boxing very seriously. Before the fight we all filed in and were looking each other over, thinking that must be the guy I'm fighting because he's the same size as me. But I won – I knocked the geezer out in sixty-four seconds!

I was flying but my trainer said, 'Charlie, you can knock mugs out all day but the time's coming when they're not going to stand there and let you hit them.' I thought that was out of line and I was hurt, but he knew I was ready for the next level up – this was kindergarten for me.

Next, I represented London in a tournament against Holland. There, I lost on points. That was tough, but I picked myself up pretty quick and started training hard for the next tournament, which was London v. Wales.

And this was where it all came unstuck.

In the second round I was winning against this guy and I got caught. I took an unguarded right hook and went down, landed badly and hit my head on the apron outside the ropes. Blood everywhere. I ended up in hospital for a week. It wasn't that serious, just leaking blood vessels, but the way it works with the British Amateur Boxing Association, the ABA, is that if you get a head injury, they basically want you out. So they took my licence away.

But if they thought I would just lie down and accept it they had another think coming: it was my future they were tossing on the scrapheap. Boxing was my way out of where I was. I'd been fighting for that future in the ring, and I was going to fight for it out of it, too.

I had a friend and supporter in the boxing fraternity, a journalist from up north called Colin. He helped me organise a press campaign to pressurise the ABA into changing their decision. We got some articles in the papers. Then I went to Roger Bannister, the guy who'd broken the four-minute mile in 1954. He was a famous athlete but also a famous neurologist. I paid privately to have him check me out. He ran a bunch of tests and scans, and said there was nothing wrong with me. The papers picked that up, too. I was so determined to keep boxing that I was also scheming to join a club up north and fight under a false name.

As it happened, the ABA gave me my licence back, but I was on borrowed time.

When I'd recovered and got fit again I fought two more bouts. The first I won. I was under a lot of pressure because the ABA had put the word out that as soon as I lost, my licence was to be taken off me. Basically, they didn't want to know. It was tough going out to fight with those two-faced bastards sitting there, waiting for you to lose.

The next fight I had a new trainer, who had the idea of taking me pro. He'd changed my style around because in pro fighting you take it slower, while in amateur boxing you just get in there and have a row. I was never used to planning things out but in this fight I'm doing what he says for the first two rounds, pacing myself, and then in the third round he says to me, 'Alright, Charlie, you need to step it up now.' I was like, 'Flippin' hell!' I'd spent two rounds keeping out of the way and now I had one round to sort it out! The bell goes and of course the other guy is right up on his toes and he wins the fight.

They took my licence away for a year. Officially it was to give enough time for any issues with the injury to show up and be resolved and whatnot, but the real reason was, they wanted me out. It was all political. The ABA are terrified of people wanting to shut amateur boxing down so if you get any sort of injury that looks bad there is no bargepole long enough to handle you. And from their experience most people didn't hang about for the year. They normally drift off into other things.

Backing down wasn't my style, however, so I said to hell with amateur boxing, I'm going pro.

I'd joined a new club by then, a pro club, where the leading figures were the Mancini brothers, Len and Dennie. Dennie took me under his wing. In the boxing world he was a famous cornerman (the guy you see frantically fixing the boxer's cuts and helping him focus and prepare for the next round). A couple of years before he'd been in the corner for Johnny Famechon, the Australian, when Johnny took the world featherweight title from Cuba's José Legra at the Royal Albert Hall. As a trainer I had Terry Spinks, who'd won a gold medal in the flyweight division in the 1956 Olympics, and had been British featherweight champion. I was the young kid in the club and I was thinking, 'Boy, what a life this is! All these famous guys around! And pro boxers, making a few quid doing what we loved!'

I had arrived.

I COULD HAVE BEEN A CONTENDER

When you're the new kid on the block you're the best thing since sliced bread. Dennie sat me down in his office and went through all the things we were going to do — the training regimes, the career strategies. He'd call me up and ask, you know, was I okay? Was I eating right? Was I training too hard? There was nothing he wouldn't do — introducing me to people, getting me deals on gloves and equipment, the lot. I felt like a star.

Work vanished from my mind during that time. I'd finished my apprenticeship by then and was doing my own jobs, but I was on autopilot: boxing was everything. Plumbing was the thing I would do while my pro career bedded in.

A few weeks before my first professional fight the year's suspension ended and my licence was up for review before the British Board of Boxing Control. The fight was going to be in Bedford and the posters with my name on them had already been printed. I remember this like it was yesterday. It was around seven in the evening and I was sparring in the gym. My trainer, Terry Spinks, came out of the office and said, 'Charlie, I'm sorry, mate. I've just been on the phone with them. They've turned you down on account of your amateur health record.'

I just kept on sparring because I didn't want to come down out of the ring.

Eventually I stopped and started phoning round for Dennie Mancini. He worked in a pub and a sporting goods shop. I finally got him at his house. 'There's got to be a way round this, and Dennie will know what to do — what hoops we need to jump through, who to kick up a fuss with,' I thought. But all he said was, 'That's too bad, Charlie. Maybe it's for the best.' He had

BOG-STANDARD BUSINESS

me off the phone in sixty seconds, his hands washed clean of the whole thing.

The pain of that is still fresh. 'You horrible bastard,' I thought, 'dropping me just like that.' The awful truth was, I was a piece of machinery with a dodgy warranty. He'd been around long enough to know what battles to pick and which ones to avoid, and there was plenty of other talent around to back. The general talk around at the time was that there were far more great fighters outside the ring than inside – because they couldn't get a licence. Now I was to be one of them.

The drive went out of me then. I showered, put my gear in the gym bag, took the bag home and stuffed my trophies and medals in it. Then I zipped it up and threw it in the attic. It got moved from house to house as the family grew, but for the next twenty-five years that bag stayed closed.

I was twenty-one then, and I never did another sport.

Some years after a businessman I respected, a friend of mine's father, said to me, 'Charlie, that was the best thing that ever happened to you. You know how short a boxing career is, and look at the business you've got.' It felt good to hear him say that. But then he added: 'You were good, though. I bet you could have been a British champion.'

I don't know if that's true, but it made it hurt all over again.

CHAPTER EIGHT

PLAN B

So my pro boxing dream evaporated on a chilly morning in 1973. I had a wife, a small baby and bills to pay.

For some people, the drive for fame and glory takes over completely, and I shudder to think of all the boxers, actors, singers, footballers and whatnot who, like me, sort of, almost, made it. How many end up feeling lost for the rest of their lives? How many end up poor and a burden to themselves and others because, guess what, there was no plan B?

I had a plan B, thank God. To me it stands for 'Plan Bill': Learn a trade, never be out of work, makes lots of money.

With boxing ripped away from me, I may have moped around for a day, maybe two, feeling sorry for myself. But I get bored easily and I soon got bored of that. And when it came time to start being the next sort of person I was meant to be, I didn't hang about.

BOG-STANDARD BUSINESS

I could have done other things, stupider things. In 1973 there were plenty of other directions hot-headed kids could take. It was party time, remember: the Age of Aquarius. Plenty I knew were off their heads on drugs, tearing around in hippie vans and hitchhiking back and forth to India. Here I want to thank my wife, Lynda, for keeping me on the straight and narrow. Nowadays having a wife and baby and bills to pay at twenty-one is seen as tragic. Most parents pray their kids will avoid all that until they're in their thirties! Even back then getting married at nineteen was a little unusual, but it was the making of me.

Lynda always stood by me with the boxing. But she was worried for me, especially after the head injury. She was even more worried when I left amateur behind and entered the pro world because she thought it was more dangerous, and she was right. But the man she'd met and married was a boxer, and she accepted that. When all that was finished, the mist cleared and for the first time I could see the needs of other people apart from myself. How many twenty-one-year-olds today, back then or any time, get out of bed if they don't absolutely have to? When my mist cleared all I could see was my pretty young wife, tired after a night with a tetchy baby, holding our little miracle, Scott, and mustering the strength to give me a smile.

I also took a hard look at our situation. We were living in three rented rooms in a house on the Old Kent Road – kitchen, bedroom, sitting room. There was an outside toilet, but no bathroom. The fact that we had no bathroom hadn't bothered me up to then because I was at the gym six or seven nights a week and I could shower there, but Lynda and Scott had to go up the road to her mother's. She and I had always talked about living somewhere nice, in a detached house of our own with a big garden for the kids, and there we were, living the way my parents lived in Camden Town in the 1950s.

It was a jolt!

I said before that I'm no dabbler – for me it's all or nothing. So I threw everything into the plumbing. I went mad on work. Not in any organised way at first. I didn't set out to build the UK's biggest plumbing business, I just worked hard at the only other thing, apart from boxing, that I knew how to do and was good at.

If I had any ambition in those early years it was to be like Bill Ellis and make as much money as a solo, jobbing plumber could make. For the previous two years I'd done plumbing as an afterthought, drifting from job to job. At one point during my pro training I'd stopped altogether and spent a few months painting, because one of the trainers had a painting company and he offered a decent wage – even though it was dull as ditch water and I hated being bossed around. But now I was back, and eager to make a proper go of it.

The boxing bubble may have popped but it left me with two very important things. For one, I'm willing to bet I was the fittest plumber in London. You need a certain level of fitness to be a tradesman, but you see a lot of lurching from task to task, heaving a big sigh every time they have to bend down or pick up a tool. Not me. I was featherweight, wiry, and flexible; there wasn't a muscle that hadn't been worked on. Stairs I took three at a time, loaded with gear, at a sprint. I'd have the van unloaded and the job set up before the next guy had said 'Good morning'. I thought four, five steps ahead, wolfed sandwiches on the go, and had the stamina to maintain that eight, nine, ten hours a day. Saving minutes here and minutes there I could fit in three or four jobs a day while anyone else would be chuffed with two.

The other thing was mental agility, being able to banish distracting thoughts and feelings. When you're in the ring you learn to choose what to focus on. Your left eye, which is closing up? The ligament that feels torn? The sinking feeling that you're points down? No, you focus on the next opportunity, and on being ready to take it.

That discipline comes in handy when you're trying to come to grips with the job at hand, calculating the time, cost, parts and materials it will require, plus juggling the rest of the day's priorities, being courteous and direct with the customer and figuring out how to please the next – all while turning a profit, day after day. Over time I got very good at that. I miss it sometimes – it feels like a dance when you're in the groove.

Meanwhile, my family was growing – our beautiful son, Samm, was born in 1976. Fortunately, my reputation and customer base was growing too. That was easy, if I'm honest. If you were around in the 1970s, you'd know why. Britain at the time was a joke. Inflation was skyrocketing, unemployment soaring and the economy was on its knees. These days you hear about countries like Greece and Ireland needing to be bailed out. Well, back then it was Britain. In 1976 we had to go begging to the IMF for a £2 billion loan just to keep the country running.

Behind it all was this absolute arrogance of the unions. Now, I believe in a fair deal for workers, but back then union power had got totally out of control. It seemed to infect everybody with this greedy, something-for-nothing attitude. If you needed your boiler fixed, for instance, most people would call the Gas Board, as it was called then. (It's British Gas these days.) And, having lodged a call with some snotty operator, it was anybody's guess whether an engineer would turn up on time, or turn up at all, or finish the job once they did show up. It's sad to say, but this attitude rubbed off on other tradespeople and the service sector in general. The British public felt they were being held over a barrel by a bunch of rude, ignorant, lazy bastards.

With all this going on, people were blown away if you turned up on time, did the job, did it fast, did it right, for the price you'd quoted, and with a smile to boot. Getting and keeping customers was a piece of cake! It's just common sense but the thing about common sense is, it ain't that common.

It started to pay off. After Samm was born I had enough money coming in to convince the bank to give me a 100 per cent mortgage on a house out in Lee, south of Lewisham. Lee and that area was the first suburban stop for people like me from the Old Kent Road starting to make good. I'd found this little Victorian terrace house for £9,000, which sounds cheap but it needed complete re-doing. Talk about creating a rod for my own back! New windows, new floors, new stairs – no small job that because it was a tall, skinny house over three storeys – and a new roof. It's a good thing I was so energetic and focussed. Fortunately I knew a lot of tradesmen so I could swap jobs with them, a bit of plumbing for some joinery or roofing, which made a big difference because if you want a job doing proper and fast, you need the right person for it. Still, I was working all hours as a plumber and then doing the same on the house.

BOG-STANDARD BUSINESS

But it was incredible! Our own house! There was only one other guy I knew who had a house out that way, and that was the boxing coach with the painting-and-decorating business I'd worked for. It was unheard of. Haunted by Camden Town, and being a plumber to boot, I made sure Lynda, Scott and Samm had the best bathroom I could possibly afford.

By the end of the 1970s I'd zeroed in on the Pimlico area with the plumbing, for a number of reasons. One is that Pimlico is posh. If you don't know it, it's the area between Victoria station and the Thames, and it's full of big old buildings and squares, and people with money tend to live there. It sounds crude, but people in posh houses need to bathe and wash and use the loo as much as those in a council estate in the Elephant & Castle, but I know who I'd rather be invoicing!

Another reason was that I'd hooked up with a carpenter called Patrick Fox, and we were mates with an estate agent called John Harper, who ran a successful outfit called Pimlico Properties. Now, any tradesman worth his salt knows that making friends with an estate agent is an open invite to a banquet of work. That's because people trying to sell their houses want work doing, and buyers moving into their new houses want work doing, and smooth-talking John Harper was happy to help his stressed-out buyers and sellers with the numbers of good people he knew. All this work in Pimlico earned me a nickname: the Pimlico Plumber.

Things ticked along nicely until January 1979, when the unions went mad. The papers called it the Winter of Discontent because the whole country had gone to the dogs. You had unions from one sector copying unions from another, demanding

PLAN B

pay rises of up to 40 per cent. Petrol stations closed because lorry drivers wouldn't drive oil tankers. Hospitals closed because nurses went on strike. Farmers were throwing dead pigs and chickens at union offices because picketers blocked feed deliveries. In Liverpool, dead people were being stacked in factories because the gravediggers had downed tools. There were rats running around mountains of rubbish in Leicester Square because the bin men refused to work. All of London started to feel like the Rockingham Estate.

My work dried up because everyone went into panic mode. The economy was going down the drain and the government – led by that dozy Labour prick, Jim Callaghan – kept talking about a State of Emergency. The only thing that kept me going was sorting out frozen pipes because, on top of everything else, it was the coldest winter in decades.

I thought, 'How can they do this? What right do these people have to bring a country to its knees?' My kids were six and three. What kind of future was there for them?

Politics had never interested me much but when Margaret Thatcher started campaigning for the election that spring I was her biggest fan. It may sound weird, a bloke like me being a fan of a posh-sounding bird like her, but she was no toff. Her dad was a family man and a grocer; she'd got to where she was by hard work and determination. And she was the only one making any sense at the time. She was against privilege of any sort, whether the privilege of aristocrats or the privilege of fat civil servants or corrupt union bosses. When she said 'Labour isn't working', she was talking to me. When she said anyone should be able to get ahead by hard work, she was talking to me. And she — a woman! — was the only one with the balls to curb the power of the unions.

When she won the election in May 1979 she set about doing everything she had promised she would, and you could hear hard-working people around the country breathe a sigh of relief. A once-in-a-lifetime leader, she brought us back from the brink and saved us from ourselves. When she died in April 2013 I had her portrait emblazoned on the side of every one of our vehicles, all 160 of them, including my old Morris and the 1950s Fordston. It was the only way I could think of to show my respect for the Iron Lady.

The fact that I even had 160 vehicles — and the largest independent service company in London — I put down to her. That's because the Britain she fought for had a place for me and people like me in it. Mrs Thatcher revived the spirit of entrepreneurialism and slapped away the dead hand of the state and the unions. She showed us we could get off the treadmill and build our own opportunities; she put the 'Great' back into Great Britain.

From the start she gave us the confidence to think bigger, which is why, that same year, I established myself as a limited company. Thanks to Maggie, Pimlico Plumbers was officially born.

CHAPTER NINE

TAKE-OFF

Looking back now, it's funny how fame and celebrity played a part in my story even before I became known as 'Plumber to the Stars'. First there was the boxing, then the Roger Bannister thing, and then, in 1983, the daughter of my carpenter mate, Pat Fox – the lovely Samantha – got her picture on page three of *The Sun*.

Samantha wasn't just a pin-up girl. She'd gone to theatre school, and been in a television play, and she'd had a band and released a record all by the time she appeared in *The Sun* at the age of sixteen. She was serious about show business, and this was the break she'd been looking for.

Pat was frantic because the phone was ringing off the hook. He said to me, 'Charlie, I need to take some time off to help Sam get this off the ground.' He told me he'd be away for three weeks, and he never came back.

BOG-STANDARD BUSINESS

Things went sour for Pat in the end. He managed Samantha for years but she ended up suing him over financial irregularities. They settled out of court in 1995 but the rift between them never healed. I'd considered Pat a friend, so the whole unfortunate episode puzzles me to this day. Samantha, meanwhile, went from strength to strength as a model, pop star and TV personality... but that's their story.

The disappearance of Pat brought things to a head. Up to that point 'Pimlico Plumbers' had still just been me, working in a loose association with Pat. We had a good formula and I guess I'd become complacent. Now he was gone and I had all these juicy jobs piling up and no carpenter. So I hired one.

Then John, the estate agent, offered to rent me a room in his basement. He had quite a big office on Sussex Street in Pimlico.

'What the f*** do I want a room for?' I asked.

'For an office, Charlie,' he said, looking at me like maybe I was a bit thick.

But I wasn't stupid, I was scared. I'd just taken on an employee, and now John was saying I needed an office. Could I carry that cost? I mean, I was doing well, but this was commitment with a capital 'C'. It's the classic fear of the little guy, and it didn't help when I talked about it with the carpenter. He just whistled and said, 'Well, Charlie, I'm glad it's you and not me.'

Then the words of the Iron Lady came back to me. I could be a jobbing solo act for the rest of my life, and that would be fine, but that's all I would ever be. If I wanted to take this somewhere I had to step out of my comfort zone.

So I took the room. I was terrified! I'd wake up in the night sweating, worrying about what I'd got myself into. But the only

TAKE-OFF

way was forward now and the thing about an office is you have to put stuff in it. So I put in a phone. Then I put in this great big answering machine. It sounds stupid now, but that answering machine made me feel like a hotshot businessman.

Today, I'm proud of Pimlico Plumbers' call centre. It's the nerve centre of the business. We continually fine-tune the way it works and I'm always investing in IT and training. Our latest graduate apprentice came up through the call centre. It's my pride and joy, and my obsession with it stems from the hours I spent with that clunky old answering machine. After being out on jobs all day I'd race back to the office and most times find it jammed with messages. I'd started advertising by then and that, plus the ten years of graft I'd put in to build a reputation, plus the fact that decent tradesmen are always hard to find, meant that my phone was ringing off the hook.

Returning calls and booking appointments took hours every evening. It was thrilling, but I couldn't keep it up. I was getting home at midnight most days. Meanwhile, our third child, darling little Lucy, had been born in 1984. As a family man, I was torn. Yes, I needed to bring home the bacon, but I didn't want a repeat of my own upbringing – with parents absent, if not in body then in mind. I needed to be around more for the kids and to support Lynda, who was run off her feet.

At the same time there was no way I could take my foot off the gas with the business. By then we'd moved again. I'd sold the little house in Lee for £30,000, pocketing a tidy £21,000, and bought a semi-detached bungalow farther out into the suburbs, out in Sidcup. It was our dream come true – spacious, all on one level, big garden – but I'd paid £46,000, so there

BOG-STANDARD BUSINESS

were hefty mortgage repayments to make. By then we were also getting used to having a bit of money. We had two cars, nice clothes – we were enjoying life.

It was another classic dilemma in the life of a small business – I needed to expand, but how? There were only so many hours in my day, and so much I could charge per hour.

I needed more MEs!

The first thing I did was hire a retired schoolteacher to man the phone during the day. Ms Jones, we'll call her. I thought she was brilliant at first: very proper and reassuring.

Then I hired another plumber, and then another. I was getting used to the idea that if I played my cards right, the calls would keep coming. And I could sense the potential for my little business, now I just needed to strap on my skates and step onto the ice.

Naturally I took a few spills at first. There were now three plumbers, myself included, out doing the jobs I'd won and in my name. But the thing about these other two plumbers was, I thought they'd be just like me, but they weren't. They weren't quite as quick as me, or as efficient with time and materials, or as careful about bookkeeping, or as skilled with the customers. Also, they seemed to make stupid decisions about where they needed to be next and what they should be doing. There were glitches in jobs and problems with customers that I never had. Instead of reeling in new work in the evenings, I found I was on the phone shouting at these guys, or getting shouted at in turn by pissed-off customers.

Now I realise that this is all par for the course. The problem was not that they were bad plumbers; it was also that Pimlico

TAKE-OFF

Plumbers wasn't their business. I said before that being a good plumber is one thing, but making money at it is another. The company was my baby and I was starting to see a future for it. It was beginning to capture my imagination the way boxing did. I had the passion and the vision for how to get to where I wanted to be, while for them it was just a job.

It was the next classic dilemma in the life of a small business: these guys needed to be managed. Time to vacate comfort zones again. I had to come off the tools, stop working *in* the business and start working *on* it. So I began spending a half-day in the office. I let Ms Jones go. She was okay on the phone but she didn't know how to make money at plumbing. I did, and I had to be there to oversee things. It felt very weird at first — I need stuff to do and that had always meant tools and problems to fix right in front of my nose. Sitting at a desk I was a fish out of water.

Nevertheless, I soon got to grips with it. Talking to customers to diagnose their problem, estimating the time and cost of jobs, organising the plumbers' schedules, ordering materials and negotiating with suppliers — I began to develop a bird's-eye view of the business. The guys didn't like it much at first. Suddenly I was all over them like a rash, telling them what to do, how to do it, and when — and checking up on them. Micromanaging, they call it these days. Well, tough luck! One of the guys adjusted, one left, and I got two more in.

Soon I was spending all day in the office, honing things like pricing, marketing, and customer care. I was fascinated by it all. Occasionally I would still go out in the evenings for emergency call-outs, but my main thing was turning Pimlico Plumbers into a well-oiled machine. Any mechanic worth his salt can tell a lot about what's working in an engine and what isn't just by listening, and during this period — the mid-1980s — I was developing my ear.

The 'Pimlico Bible' started to take shape around this time. This was, and still is, a living document that guides every aspect of the business. There are not many small and medium-sized businesses out there that have a 'Bible', and you could say that it shows what a control freak I am. In this you would be absolutely correct and I make no apologies for it. I was fast learning that if you want to build something valuable and unique, something that stands out in the wasteland of mediocrity, you can't expect everybody to jump on the bandwagon, or even 'get it'.

I had rules and standards. They had worked for me for more than ten years by then, but getting my plumbers and tradesmen to take those rules seriously proved a constant, uphill battle. I found I was forever printing out notices and pinning them up around the place, from basic principles — 'Always Leave A Satisfied Customer' — to rules designed to stop problems coming back and biting us on the bum — 'Do Not Carry Out Work On Obsolete or Unfamiliar Appliances' — down to generally how to behave — 'Uniforms Must Be Clean, and Worn at All Times'.

The problem was, it made the place look messy, and I was running out of wall space. Also, once a notice had been up for a few months and became faded and hemmed in by new notices, people felt they could ignore it: the temptation to take shortcuts and fall back into old habits is very strong. They would see every new sign as Charlie's 'new thing', while to me it was another finger in the latest hole in a big, leaky dam.

TAKE-OFF

I needed a way to enshrine all this stuff, to turn it from what it seemed to them, a bunch of shifting whims, to what it actually was, the framework and philosophy of the business — in other words, the law of the land. So I ripped down all the notices and put them into a folder. This helped me see what was missing, and I added new pages. Soon it was a couple of inches thick, covering everything from personal appearance (no piercings, tattoos or extreme hairdos) and customer etiquette to job pricing, forms and complying with regulations. It was a complete 'Dos & Don'ts' reference book. I made copies and gave everybody one. A powerful tool, it allowed me to say, 'Here, this is what's expected of you. It's not rocket science. It's your job to understand it and follow it. If you don't, you're out.' It took the onus off me to badger them. A couple of guys put it to the test and I sacked them. The message stuck then.

Over the years we've added to 'The Bible' as regulations change, and to take account of new ways of doing things brought on by the Internet and mobile phones and whatnot, but the core is still the same, and people all over the world have asked me for copies so they can adapt it to their own businesses.

I started building up the Pimlico Plumbers' fleet of vans in this period, too. There was a fair amount of resistance to this: most guys want their own vans. A tradesman's van is his personal space, like a bedroom. To them it can be as messy and smelly and clapped-out as they like — it's nobody's business but theirs. Naturally I didn't see it like that. My tradesmen's vans were travelling adverts for my company. What did a rag-tag bunch of vehicles — this one beat up, that one dirty, the other with its exhaust half off — say about Pimlico Plumbers? It said we were cheap, shabby chancers like everybody else. On the other hand, what does a fleet of identical vans that are spotless and liveried say? It says we are consistent, professional, exacting and successful. Given a choice like that, who would you prefer to be in your house, fixing your plumbing? (Common sense again, and again; common sense ain't that common!)

For a while we messed about with guys putting magnetic logos on their vans and promising to keep them clean and in good repair, but things always slide. There's always the day when they can't find the logo in their garage, or when they're running late and can't be bothered with it. And in most cases their vans would sooner or later revert to their natural 'bedroom' state. It was a whole other area of conflict and high blood pressure, so I started engineering the problem out by making them drive my vans and making their employment conditional on keeping them spotless. It was expensive but if you want quality you need control, and if you want control, you must shoulder the cost.

I was getting more confident about this sort of thing because my approach was obviously working. There wasn't much time to stop and look around in the 1980s but when I did, as the decade drew to a close, I was surprised by what I saw. We had taken over the whole basement of Sussex Street. Our office had grown from a 100-square-foot room to 500 square feet – still very modest, but still a five-fold leap. Meanwhile, things were getting cramped. I had an office full of people now – people doing estimating, people doing accounts, even a rudimentary call centre. And parking for the vans was a nightmare.

It was time to move.

CHARLIE ON... HAPPY WORKERS

Getting the best out of your staff can be the difference between success or failure. Business owners set direction and strategy, but it's up to the workforce at the front line to deliver. The question is how do you reward them?

A survey released over the weekend [early May 2014] suggests that a well-timed 'thank you' from the boss feels as rewarding as a £1,600-a-year pay rise. I'm not so sure that everyone would take gratitude over cash, but it makes a good point that money isn't the only motivator.

According to the survey, a simple 'thank you' can lift workers' moods and output and even boost company profits. Now that's something I can agree with. And get the right balance between praise and pay and a business can be onto a winner.

Giving someone a good day's pay for a good day's work is at the heart of how our society operates and, despite a few shady characters still paying below the minimum wage, it's a practice that is serving our society well.

Simplistically, it gives people the opportunity to aspire to do things like rent or own their home and take holidays. These are the sort of things that get people out of bed in the morning. However, money really isn't everything and if people aren't happy in their work they won't be productive, which doesn't do anyone any favours.

There are so many ways in which employers can boost morale and 'buy-in' to company ethos beyond remuneration and, as this survey suggests, a timely thank you is one of the most straightforward ways.

BOG-STANDARD BUSINESS

Recently at Pimlico Plumbers we introduced an incentive scheme where, if the company has a good month financially, all meals and snacks in our canteen will be free for one day.

It makes good sense for the people entrusted with looking after your customers to be happy and content at work. That's why when we beat the sales total for the same month the year before, the chilli con carne, roast beef, sandwiches or whatever's on the menu will all be gratis for a day.

Communicating gratitude to staff is not only the right thing to do; it's the polite thing to do. Interestingly a fifth of the 2,500 workers questioned for the poll said their bosses were ungrateful or in some cases rude.

I can't see why employers have to act like eighteenth-century mill owners when they're interacting with their staff. It's not productive, but being polite could be one of the best non-financial investments they make, after all manners cost nothing!

Originally published on 6 May 2014 on realbusiness.co.uk – The UK's leading title for high-growth businesses and entrepreneurial SMEs.

CHAPTER TEN

INTO THE HOLE

When I was an apprentice plumber going to jobs on the bus I would regularly pass by a handsome Georgian house at 136 Lambeth Road and I never failed to stare out the window at it with a mix of awe and envy.

They were plumbers! The sign out front said 'Holborn Heating'.

I thought they must be the bee's knees, having premises like that: so posh, and so close to central London. That tucked-away part of me that still saw myself as a plumber — and not a boxer — wanted success, to be sure, but it was the Bill Ellis sort of success at that point: plenty of freedom, work and cash. These guys at 136 Lambeth Road, they were in another league altogether.

Then one day in late autumn 1990 I was driving down Lambeth Road and I noticed a for-sale sign out front. I probably hadn't even thought of the place in years or, if I had, it was only in that passing way, the faint tug of remembered feeling. But seeing that sign hit me like a ton of bricks, I pulled over and sat staring at the building just like I used to do.

There was a lot of interest in the property, but it needed complete refurbishment. I told myself it was perfect. It had 2,000 square feet of space over three floors, with garages out back that used to be stables, plus more parking in the front. It was solid, respectable and conveniently located. More than that, it was like a piece of me, my history and my future all rolled into one. Fate, you might call it.

Two hundred and thirty-six thousand was the asking price, and that's exactly what I offered.

Say hello to me, Mr Flash Git, in 1990. What I'm driving that day on Lambeth Road is a brand new, top-of-the-line Range Rover – bought with cash, thank you very much. My company was now turning over a million pounds a year. On the home front our brood had grown to four, with the joyous arrival of Alice a couple of years earlier. We'd sold our semi-detached bungalow for seventy-five grand and moved into the domicile of our dreams, a big detached house with an acre of garden, still in Sidcup, that I'd picked up for £110,000. I seem to remember it needed some work doing but it wasn't me wielding the tools anymore – I got some guys from the company to do it.

We were still in the basement of Sussex Street, Pimlico, but not for long. I'd bought a building down in Camberwell with a bit of a yard for £66,000 and we were doing that up. We'd been going to move there but now I'd spotted 136 Lambeth Road and convinced myself that I had to have it, the plan changed, or ballooned, I should say. Lambeth Road would be Pimlico Plumbers HQ and Camberwell would be our first branch office.

I was about to learn the most important lesson ever in my business life – and nearly lose my shirt in the process.

I wouldn't have said to myself that I'd 'made it'. To this day I wouldn't say that. There's a mental block there that prevents it. For people like me poverty is always following, just out of sight. But I was listening to a sneaky little voice in a corner of my head that was saying, 'Psst, Charlie, look: everything you touch turns to gold!' The business was taking off and all our houses I'd flipped for a healthy profit. As a sideline I'd even bought a newsagent's in Sidcup for £3,000. Lynda and her father ran it for a couple of years and we'd sold it for eleven grand – small change in the big scheme of things but we'd still nearly quadrupled our money. I was only thirty-eight and I felt unstoppable.

Buying Lambeth Road for over a quarter of a million was another thing altogether, though. How was I going to do that? No problem, as it turned out. My accountant put me on to this financial advisor who had all the answers. I would just get a loan. Two loans, in fact: half the cost of the property from a building society and half from a bank, Barclays. The clever thing about this, he said, was that if things went pear-shaped it would be harder for one of the lenders to force the sale of the property. I was thinking, yeah, whatever. As if that's going to happen!

I couldn't get over how easy it was. They sent a surveyor round. I made sure I was there. He was jolly and accommodating. Yes, he said, the property's worth two hundred and thirty-six grand, easily. Then the two bank managers came to see me.

BOG-STANDARD BUSINESS

Blown away by me, they were blown away by my business too. Everything was simply super, they said. I'd never met them before and all I had to do was sign a piece of paper, put up my house and the Camberwell property as security, and I had a quarter of a million pounds to play with. That was the first business bank loan I'd ever taken out and I was thinking, why hadn't I done this before?

We took possession in December 1990. I made sure I had a team of locksmiths with me when I went there for the first time as the new owner because I'd spotted this huge safe under the stairs, to which nobody had any keys. It just shows how naïve my thinking was at the time because I was convinced there would be a pile of money in there and I was seriously disappointed when it was empty!

Never mind. We got busy gutting the place. This was definitely becoming a pattern with me, something to do with my needing to put my own stamp on things. I wanted everything to be right and proper and that usually meant starting from scratch. Up came the carpets and the floors. Off came the plaster. Down came the ceilings and a few non-supporting internal walls. I wanted a complete blank slate.

Now, if I'd have been less absorbed in my own thing, I may have noticed that there was a recession going on. I mean, everybody had been talking about it, and a big dump of snow in December 1990 had brought the economy to a standstill, but I'd just kept motoring on. I couldn't take it seriously. The last recession of the early 1980s had gone straight over my head because I was too small, as a business, to be affected. Unemployment topped two million in March 1991 and I was like, so what? Business was okay, as far as I could tell, and the banks had just lent me all this money. What recession?

By April it was a clean shell and we were ready to begin.

And then came the knock on the door.

Barclays sent the surveyor round again, the same jolly fellow who'd found everything so satisfactory before. Except he was a different man this time: grim, tight-lipped, with a face like thunder. 'What's eating him,' I wondered?

'Why have you ripped out all the plaster?' he demanded to know. 'Why have you knocked down the walls? This building's no use to anybody now.' For a split second I thought, 'Aw, bless, the bloke's a bit simple.'

I started to explain: 'It's okay, we have plans; it's going to look great.'

'No, no,' he was going, 'You've fundamentally altered the value proposition of the property.'

I was like, 'value proposition?' We couldn't use it the way it was — it was a nasty old warren.

'Be that as it may,' he said, 'with the current economic conditions, it will be seen as a precipitous blah, blah, blah…'

'Mate,' I said, 'look, in business this is what you do!'

But he was having none of it: 'I'm afraid I'm going to have to alert Barclays to the situation,' he told me.

The property was now valued at around fifty grand, and I'd have to sell immediately.

My mind whirred. Inside five seconds my education about the ways of banking was complete. The penny dropped, and hard. In December I was the best thing ever, with a great track record, a splendid business, and a blinding plan for the future that would make both me and Barclays loads of money — after all the interest on the loan was an eye-watering 17 per cent.

But by April, through no fault of my own and without the fundamentals having altered — except in the bankers' lizard brains — I was the biggest wanker going. They wanted to claw their money back and they didn't give a toss about me. It wasn't the recession threatening to put me under. Business was actually still fine — I could see by then that we'd have a storm to weather, but there was no reason we couldn't come out the other end. It was the bank, racing for the lifeboat and hacking at the ropes before anyone else could get in.

I'm afraid I lost my cool at that point. My manners departed. I told him to get out. In the lingo of South London, I f***ed him off *right out of it!*

It was war then. The surveyor scuttled back to the bank, and Barclays began a non-stop campaign to get me to sell, and not just Lambeth Road, but the Camberwell property too. They wanted me to put Camberwell up for auction with a guide price of forty grand, and the proceeds, of course, would all go to them. They even wanted me to sell the Range Rover. 'Right, Charlie, what are you driving?' the Barclays wanker asked in one meeting. I was that beaten down that I told him. 'Okay, I want you to sell it, this week, for cash. I want you to fax me through the valuations right away.'

When I didn't jump to their orders fast enough they put the squeeze on. I had an overdraft for £80,000. In another meeting the wanker said he was dropping it to seventy.

'You can't do that,' I said.

'I just have,' he told me.

INTO THE HOLE

Every month it would come down by another £5,000 or £10,000. As long as I was making the monthly payments, this was all they could do, and they had no qualms whatsoever. It was a simple strategy: ruin my business so I had no choice but to roll over and die. Like tying my hands behind my back and pushing me into the deep end.

'We have no choice, Charlie,' they'd say, 'we have to limit our exposure.'

'Why don't you just f***ing take the lot?' I'd say.

'Well, okay, we will, then, if that's the way you feel,' they'd say.

And then I'd have to ring them up and beg and plead.

I did sell the Range Rover but I didn't tell the bank, and I used the money to keep the business going. The manager was hopping mad when he found out.

Things hit bottom for me around October 1991. I remember on a Friday, just as we were laying out the paycheques, the bank called and told me they were cutting the overdraft from £20,000 down to £10,000. A lot of staff had sensed the ship was sinking and had left, but I still had around twenty people working for me then. I felt completely alone. My accountant and my financial advisor were always on at me to 'work with' the bank, even though I knew it meant the end of the business. Later, I found out that they all knew each other, them and the Barclays' wankers. I'd tried to keep Lynda out of it, but she knew what was going on and was terrified of losing the house. Meanwhile, the hole I was in was getting deeper. With any spare cash going on keeping the business afloat, I'd fallen behind with tax, national insurance and payments to suppliers.

BOG-STANDARD BUSINESS

That, plus the bank loan put me in debt to the tune of half a million pounds.

Desperate for some new course of action and sick of the people around me, I sacked my accountant and told him to take the financial advisor with him. On a recommendation from a friend I got a new accountant, a straight-talker called Bernard Gross. In my first meeting with him I told him I wanted to go bust.

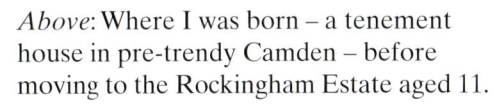

Above: Where I was born – a tenement house in pre-trendy Camden – before moving to the Rockingham Estate aged 11.

Below left: With my Dad Sid in around 1980.

Below right: On a night out at Pontins with my wife Lynda when I was 23.

MULLINS HALTS GALLANT WALLY

CITY —October 5

BANTAMWEIGHTS Charlie Mullins (Fisher) and Wally Angliss (Fitzroy Lodge) provided a stirring punchup on the Stock Exchange dinner show at Dunster House. Mullins was named the winner at the end of the second round when the referee stopped the bout because of damage under the right eye of Angliss.

Both had refused to give ground in the first, when sustained attacking and toe-to-toe rallies had the crowd roaring. The big punching was coming from 19-year-old Mullins, who ripped in some tremendous left hooks and right hands, but brave Angliss retaliated with flurries of hooks to head and body.

The second round was fought at a fierce pace with first Mullins then Angliss getting on top. Mullins scored with heavy left hooks while Angliss was having difficulty in getting through against the elusive, bobbing and weaving tactics of the Fisher boy.

Nearing the end of the second, with Mullins appearing to be getting on top, Angliss staged a brave rally after being shook-up with combination punches. He bravely backed Mullns on to the ropes, but again Mullins bobbed and weaved his way out of trouble, crashing in some fair left hook counters.

At the end of the second, with the right eye of gallant Angliss grazed and swollen the referee stopped what had been a thriller. A return is a must.

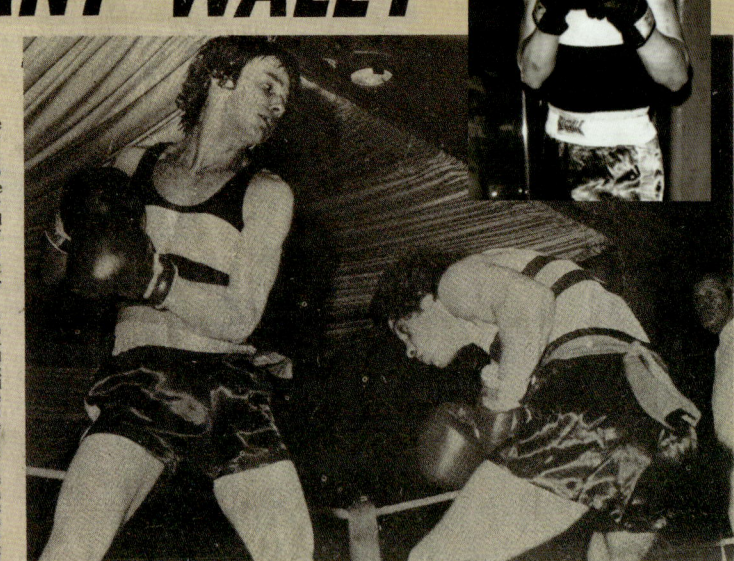

ANGLISS ducks under fierce left hook from Mullins.

Above: With my boxing teammates in 1970. I am on the left of the front row.

Below: Extract from *Boxing News* in October 1972.

Inset: Me in my heyday.

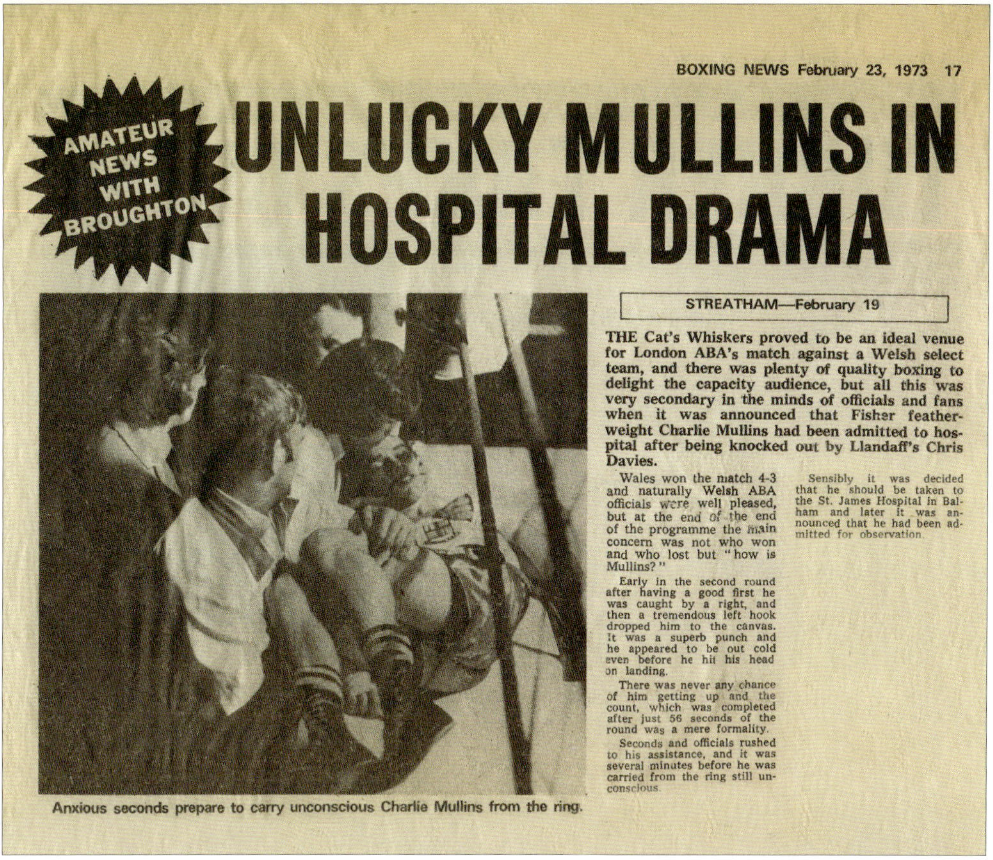

BOXING NEWS February 23, 1973

AMATEUR NEWS WITH BROUGHTON

UNLUCKY MULLINS IN HOSPITAL DRAMA

STREATHAM—February 19

THE Cat's Whiskers proved to be an ideal venue for London ABA's match against a Welsh select team, and there was plenty of quality boxing to delight the capacity audience, but all this was very secondary in the minds of officials and fans when it was announced that Fisher featherweight Charlie Mullins had been admitted to hospital after being knocked out by Llandaff's Chris Davies.

Wales won the match 4-3 and naturally Welsh ABA officials were well pleased, but at the end of the programme the main concern was not who won and who lost but "how is Mullins?"

Early in the second round after having a good first he was caught by a right, and then a tremendous left hook dropped him to the canvas. It was a superb punch and he appeared to be out cold even before he hit his head on landing.

There was never any chance of him getting up and the count, which was completed after just 56 seconds of the round was a mere formality.

Seconds and officials rushed to his assistance, and it was several minutes before he was carried from the ring still unconscious.

Sensibly it was decided that he should be taken to the St. James Hospital in Balham and later it was announced that he had been admitted for observation.

Anxious seconds prepare to carry unconscious Charlie Mullins from the ring.

Above: The hospital drama that ended in me losing my licence.

Below left: With my son Scott in 1972.

Below right: Feeding the horses on holiday with baby Lucy in 1978.

Above: When the business was taking off: Lambeth Road in 1994.

Below left: Taking a call outside the Lambeth Road premises.

Below right: Today: Me and the Bentley outside our Sail Street offices.

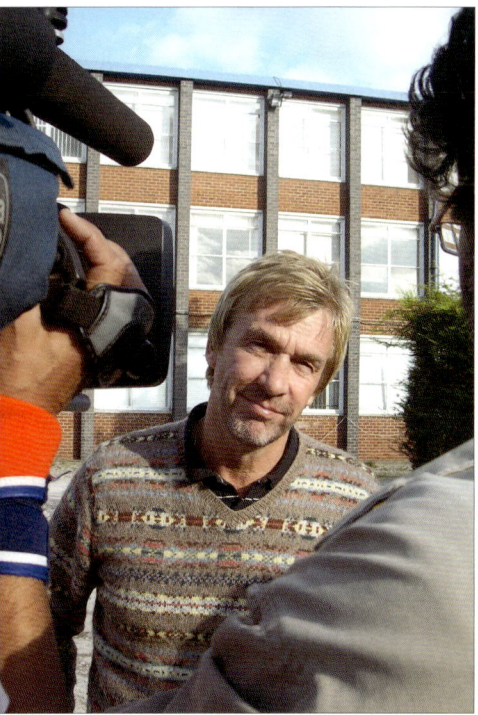

Above: A PR hit – the personalised number plates have gained us huge popularity.

Below left: Filming *The Secret Millionaire* in Warrington in 2009.

Below right: Buster took up running in 2007, and went on to race the 10km Bupa Great Capital Run, the Roding Valley Half Marathon and the London Marathan aged 101.

Above: Charlie meets Charlie: at a function for The Prince's Trust in Buckingham Palace in 2010.

Below: In September 2008 the Capital Radio breakfast show was broadcast from Sail Street for a winning bid of £50,000 in a charity auction of Help a London Child. Here I am alongside presenters Johnny Vaughan and Lisa Snowdon and dancer Brendan Cole.

Above: Boris Johnson pays the offices a visit during the 2008 London mayoral race.

Below left: Posing alongside David Cameron at the Conservative party conference in Birmingham in 2014 where Pimlico Plumbers had a stand focussed on apprenticeships.

Below right: George Osborne at Pimlico Plumbers during the 2010 general election campaign.

Above left: Arriving at an event for business leaders in 2013 – one of many visits to Number 10.

Above right: Travelling in style – it's amazing the welcome I get when I turn up in this.

Below: With some of the Pimlico Plumbers team during the filming of Channel 4's *Show Me Your Money* documentary in 2012.

CHAPTER ELEVEN

DIGGING IN, AND DIGGING OUT

Bernard wasn't convinced, but he said okay. He called in a favour and got us in to see this famous liquidator, a guy called Panos Eliades, who had a posh office up in Bloomsbury.

Now here I have to relate a bizarre and, at the time, painful coincidence. This hotshot insolvency dude, Eliades, had that same year begun to promote the heavyweight boxer Lennox Lewis, who'd won a gold medal in the 1988 Seoul Olympics, plus the European and British heavyweight titles. The following year Lewis would knock out Donovan 'Razor' Ruddock, making him the number one contender for the World Heavyweight Championship title.

Desolation was all I felt as we filed past guys queuing up outside his office and all the way down the block, waiting their turn to go bankrupt. Here was Eliades all over the news for promoting the next rising star in boxing. And me? Shut out

BOG-STANDARD BUSINESS

of boxing, I'd turned to business and done well for a while but now I was here about to ask him to put a bullet in my company's head.

Eliades was brisk and businesslike. He sat us down at this huge table, itself about the size of my living room.

'This doesn't need to take long,' he said. 'I have two questions. First, how much do you owe?'

We told him: around half a million.

'Okay,' he said, 'and how much have you got?'

We told him: there's the Camberwell property and, aside from that, sweet f*** all.

He said, 'Well you don't have to have much of a brain in your head to work out what you need to do.'

Go bust, he meant.

Three days was all the paperwork would take, and in three days I would be back up and trading under a slightly different name. It would cost £3,000. Job done.

I'm ashamed now to admit how relieved I felt: it could all be over and I could start again with a clean slate. But walking out of there and past the long line of waiting bankrupts, Bernard, my accountant, was unhappy.

'I don't like it,' he said. 'It wasn't that long ago you were turning over a million pounds. Your business is still good, your reputation is still good. I think we should get a second opinion.'

So we did. We went to this other guy Bernard knew and he had a different take on things. He wasn't in such a hurry to hustle us out the door three grand lighter and he pointed out something Eliades had glossed over: if I went bust and defaulted on the loan, I'd lose the house. Simple as that. I would have

DIGGING IN, AND DIGGING OUT

to move Lynda, Samm, Lucy and Alice out and into, what? A council flat? Samm and Lucy were doing their exams. Alice was eight. Scott was travelling – he'd have nowhere to come home to. I mean, strictly speaking, we didn't need all that space, but the house was a symbol for everything I'd toiled for these last twenty years.

Other things dawned on me. I didn't give a stuff about the bank but if I walked away from my other debts there would always be people – suppliers, staff – whom I'd cross the road to avoid. And it wouldn't be Pimlico Plumbers anymore. The name would have to be different – Plumbers of Pimlico, or Pimlico Heating and Plumbing, or something else that was equally not the real thing. People would know I'd gone bust. More than that, *I* would know.

Something came together inside for me then. I thanked the guy and walked out of his office, determined to get back in the ring. Bernard was great. 'Charlie,' he said to me, all happy now, 'if you come out of this, and I think you will, the business is going to be stronger for it.'

Companies can grow fast but they don't always grow true. I knew from my boxing career that when you hit a ceiling in your performance it's time to unlearn some bad habits. So I took a hard look at Pimlico Plumbers and the things I was doing, and started making some changes.

We'd never really had a proper business plan because I'd always just sort of made it up as I went along. One of the results of that, as Bernard helped me to see, was that I had too many wrong people in important jobs. Up to then I'd get a mate to do something, or switch somebody over to a role who fancied he or she could do it, and then I would meddle and interfere when I wasn't happy. It meant I was running around fighting fires all the time. I needed to get the right people in for the job and then step back and let them do it.

The first to go was the accounts manager. I'd made him accounts manager because I liked him and he seemed careful with his own money so I thought he'd be careful with mine, but he had no qualifications or experience and didn't really know what he was doing. Bernard told me the sort of person I needed and the sort of things this person should do, so, in a grown-up way, I advertised the job properly and interviewed people. When I'd got the new guy in it was blindingly obvious how much the other one had missed and what hard work he'd made of everything. I did the same thing with the HR manager.

The next thing to sort out was cash flow. Like most businesses we'd always invoiced customers and allowed them to rack up bills on account, which meant we were forever chasing payment. At one point we totted up all we were owed and it came to eighty grand! As Bernard said, 'Neither a borrower nor a lender be.' It was hitting home to me that the people who go under in a recession are the ones who owe money and those who are owed money so we stopped serving customers on credit. Overnight we switched to payment on completion, and there were to be no exceptions. If the Queen called us out and wasn't prepared to pay by cash, card or cheque, right then and there, we wouldn't do the job.

It wasn't easy making this change. We had to write it into the script for the people in the call centre and watch them to make sure they were getting the message across in a firm,

DIGGING IN, AND DIGGING OUT

polite way — and getting agreement from the customers. The tradesmen had to get tough without losing any of their courtesy or professionalism. To be all smiles and then to say, 'I've finished the job, here's the bill, can you please pay me now?' was a big adjustment! A few didn't take it seriously and we had to make examples by giving them the sack. Eventually the message sunk in: if you ain't coming back with the money, don't come back at all!

How I wished we'd done this from the very beginning. (The funny thing is, they did this at F&H Plumbers, down in Loughborough Junction, where I'd finished off my apprenticeship. Oh well, I guess some lessons take twenty years to learn!)

All of a sudden we had all this cash coming in. And the jobs went a lot more smoothly. If a customer knows they've got to pay on completion, they engage more and police the job better. They won't part with their cash until they know the boiler works, for instance, and feel sure they understand what's been done. Complaints and callbacks all but disappeared.

At the same time we implemented a new pay structure for the tradesmen that allowed them to make more money, the more happy customers they had. Before, when a customer called, we would allocate the first available Pimlico Plumber, and the customer had to take whoever they got. Now we encouraged the plumbers to leave their business cards, and allowed customers to request the plumber they wanted. In that situation, you really see how the cream rises. Plumbers who were good and polite and went the extra mile started getting popular. And the more popular they were, the more hours they racked up. A plumber who was acceptable could trundle from job to job assigned by the call centre and put in a steady eight-hour day. A hungry all-star would have no trouble working twelve or thirteen hours if he wanted – and have the pay packet to show for it.

We also firmed up our twenty-four-hour service offering, basically by making sure we had people available to take calls and respond to them around the clock.

Lynda joined the business during this time, working in the call centre, and that was a massive help. You need somebody in your corner when you're shifting people out of their comfort zones and making unpopular decisions. I've always been seen as a hard bastard but the truth is, back then, it was too often a case of the tail wagging the dog.

In 1992 I was finally able to get Barclays off my back. I'd sold the lease on the basement of Sussex Street, did a basic, temporary refurb on Lambeth Road to get everybody in, and sold the Camberwell property for £90,000 – a lot better than the £40,000 the Barclays wankers had insisted on the year before! The profit from that, plus some extra I managed to scrape together from God knows where, allowed me once and for all to get shot of them. It's interesting that throughout all this the building society who lent me the other half of the loan had been honourable. They hadn't pressed the panic button or tried to put me out of business.

Immediately after, I closed my Barclays account and went to another bank. I pledged never to have anything to do with Barclays again. They'd set out to destroy me and would do it again at the drop of a hat. Needless to say I've never had another bank loan and I'd have nothing to do with banks full-

DIGGING IN, AND DIGGING OUT

stop if I didn't need to. If I could keep my money under a mattress, I would.

I've been with my current bank for twenty-two years and have never met any of the fifty or so managers they've had in. They keep calling me up, though, and the conversation usually goes like this:

'Hello, Mr Mullins, we'd like to come and see you.'

'No.'

'It's just a courtesy visit to see how your business is doing.'

'Look at my account. That's how my business is doing.'

'Well, we'd love to see what other services we can offer you.'

'There is nothing you can offer. Goodbye.'

'Are you sure we couldn't pop round for a quick chat?'

'If you did pop round, you would be ushered off the property without delay and with even less ceremony.'

Bankers: crooks in suits.

CHAPTER TWELVE

FINE TUNING

We weren't out of the rapids yet. For the first half of the nineties it was touch and go but by 1995, when people started spending money again and were looking around to see who they could spend it with, so many companies had gone bust. Many good people had retired early or left the industry altogether. But there we were, good old sturdy Pimlico Plumbers, having weathered the storm. It sent shivers down my spine, thinking they had no idea how close we'd come to abandoning ship.

And Bernard had been right: we were stronger for it. We had a solid structure in place, with the right people in the right places, doing the right things. Our Lambeth Road headquarters, which Barclays had tried – and failed – to snatch from us, had now taken definite shape. On the first floor we had a reception area where we were starting to build up our collection of awards and celebrity testimonials. Behind that we had a general office for admin and accounts. On the second floor we'd knocked two rooms into one to make a call centre, and on the top floor was my office and the estimating department. It worked beautifully, and the size and central position of the call centre made me feel proud. The stress and panic of 1991–92 was still very fresh in my mind though, and sometimes I would just stop and listen to the phones ringing and the hubbub of the operators below, and I would tell myself, 'Charlie, it's okay, you're in demand.'

We'd had a quarter of a million pounds knocked off our turnover in the recession but it was on the move upward again now. I think of the early nineties as the resurrection of Pimlico Plumbers, or better still, the foundation of the 'real' Pimlico Plumbers.

There's always a reason not to bother with something you know you should maybe bother with. On the one hand there's never enough ready time or money, or know-how to get down with the slog. Then again, the secret of success is bothering to put in the slog and expense to lay the right foundations for later. I am extremely lucky in this regard because I'm a control freak, a busybody and I hate getting ripped off.

I mentioned before that we'd begun building up our own fleet of vans so that we could control the image of the company out on the road. Well, by 1995 we had around twenty vans, which was great, but also brought a problem of its own: we were spending a fortune in local garages on repair and upkeep.

My attitude to vans is quite straightforward. I may have spent a lifetime distancing myself from the image of tradesmen as 'cowboys', but I do see the plumber's van in exactly the same way as a cowboy sees his horse. First, just as a cowboy ain't a cowboy without a horse, so is a plumber anything but a plumber without a van. Second, a cowboy puts his horse before himself. When he rides into town after weeks on the trail, no matter how tired, hungry and desperate for a drink he is, he makes sure his horse is watered, fed and bedded down before he sorts himself out. The reason for this is simple: he doesn't know how the evening is going to pan out, and he can't leave town if his horse ain't fit and ready.

I've always insisted our vans be clean and in top working order at all times because what does it say about a plumber if he calls you up and says, 'Sorry, mate, can't come, the van's broken down'? It says he's a dickhead, that's what! So at the slightest sign of trouble our vans went into the garage but this put us completely at the mercy of the garage.

I've got nothing against garages, but they've got their own priorities and their own systems, and they need to make as much money out of you as possible. How often do you use a garage? Two or three times a year? How often do you come away saying, wow, that was fast, convenient and cheap? Now multiply that by twenty or thirty times and you'll see what a headache and rip-off it was starting to feel like. At any one time it seemed I had more vans in the garage, with the people there stalling and messing me about, than I had on the road. So I hired a mechanic and kitted out the garage behind the office with everything he needed to keep my guys mobile. What a difference that made! It wasn't cheap, but overnight we regained control of a crucial part of the business.

By 1995 we had enough vans that parking was becoming a

BOG-STANDARD BUSINESS

nightmare again, so I took out a lease on a railway arch just around the corner, giving me an extra 3,000 square feet of parking space. It meant Pimlico Plumbers' footprint on Earth was now 5,000 square feet – puny, I know, in the scheme of things, but ten times what it was in the basement of Sussex Street. *Ten* times! I started to get excited.

During this period we matured in other ways, too. We started to develop a distinct look, and for this I have to thank the best graphic designer I know, Tony Davidson. Tony ran a small chain of print and design shops, and I started using him for ads and flyers at the end of the eighties. I used to lay out all that stuff myself but even I could see it needed a professional's touch. Tony was good: whether it was an ad or a flyer, he would listen to what I wanted, and pretend to respect the efforts I'd made, and then go and do exactly the right thing! But I was such a fusspot and always changing my mind – control freak, in other words – that he was spending more and more of his time on my stuff, and less and less on anybody else's. He did a lot of work for the big corporates and he could charge top whack for that, but the downside was he'd do the work and then wait weeks for a decision, only to be told, sorry, no, they'd be sticking with the old logo after all. I was a pain in the backside but at least when I came to him something emerged as a result.

He took a hit in the 1991–92 recession, as we all did, and one day he said to me, 'Charlie, I seem to be doing most of my work for you these days. Why don't I join you?'

I thought that was a great idea, so he shut his shops down and came on board. He's my marketing director now.

Tony designed the first of our new logos. This kept me up

nights worrying because it seemed so, I don't know, unusual. I liked it a lot, but I had real trouble with the fact that you just didn't see any other plumbing firms with logos like that.

'Exactly,' Tony said. 'This will get you recognised. And if you start getting recognised for the right reasons, your business is going to grow.'

Once again I had to let go and trust the right person, and I'm very glad I did. That logo was a hit, and really helped us to stand out.

Tony redesigned our uniforms, too. I had always insisted on uniforms and it still wasn't all that popular amongst the tradesmen and engineers, but Tony helped by revamping the old dark-coloured bib-and-brace overall to a more modern style with bright colours. The guys took to them better, and I was pleased because someone turning up in a clean, bright uniform gives a strong impression to customers. For one thing, it suggests they're happy in their work. Customers like people who are happy in their work, they don't want some surly misery-guts in their space. For one thing it's unpleasant. For another, they worry the guy will do something underhand out of spite. And when you can get your workers to *act* like they're happy, you've got a winning combination.

This was one area of the business I found I just could not let go of — recruiting the tradesmen and making sure they looked and behaved how I wanted. It was so clear to me by then what our trademark was: courtesy, professionalism, and transparency. If a person calls a plumber, something has gone wrong with his/her home. They're feeling embarrassed, stressed, and vulnerable. They're worried they'll be robbed blind, kicked while they're down. What we offer with our skill is a solution to the problem, and also reassurance, by the way we behave. If that customer doesn't call us when the next problem hits, we've failed.

It seems so obvious — painfully obvious — but developing a workforce that really gets it is far from easy. You need the right people to start with, people who can communicate in clear English with a customer, with the office, and with suppliers. They must be house-trained — that is, know how to behave, remember to take off their shoes, put down a dust sheet, respect people's furniture and privacy, and a hundred other things. That's why we prefer our tradesmen to be at least twenty-five because there's a good chance they're married or in a relationship by then, and wives and girlfriends are essential for house-training. They pick up where mums have to leave off!

The sad fact is it's hard to find people like this in the trades. In other countries, like Germany, skilled and semi-skilled tradespeople are respected, as they should be. But in this country there's the class thing, and on top of that we tore down the old vocational training and apprenticeship system that used to offer a decent place in society for those who weren't cut out for university, or couldn't afford it. Denied respect, people don't respect themselves or others, and they don't know how to be in society other than chippy, hopeless or aggressive.

Still, there are good people out there if you know how to find them.

I can tell in sixty seconds if someone's not right. You'd be amazed how many turn up with fingers bright orange with nicotine, filthy nails, unshaven, dressed shabbily. And the number of times they make an excuse about why they look like that! 'Oh, I had to change the carburettor…' It just makes it worse — it means they know they should have made the effort and they're willing to lie about why they didn't.

It took a long time before I could hand this over to the HR department. The first time I tried to teach a new HR guy this I let him do an interview and told him to make a decision within five minutes. 'If he's wrong, get rid of him,' I said. 'If he seems alright, go a bit longer. It ain't complicated, just ask yourself, are you comfortable sitting there with him? Would you want this geezer in your house?' I left him to it.

Fifteen minutes later he came to my office, saying, 'Charlie, I just don't know. Can you come and talk to him?'

So I went in. He was wearing a leather jacket. Bzzzt! went my internal buzzer. You don't turn up to an interview for a public-facing service job in the uniform of a biker. 'Excuse me,' I said, and leaned down and reached across him to pick up a pen from the other side of the desk. It was awkward and he leaned back, but not before I'd caught a whiff of booze, which was what I was checking for. I'd have ended it there but I was curious. 'So where you working now?' I said. I think he caught my drift because he shifted nervously.

'I'm, ah, managing a pub at the minute,' he said, giving his face a rub-down. 'Just temporary.'

'Right, okay, thanks,' I said, and left.

'Get rid of him,' I told my guy. The decision was made in twenty seconds — ten, if you don't count the supplementary question.

'What's wrong with him?' the HR guy asked, all agog.

I explained about the booze. 'Pub or no pub,' I said, 'the guy's

BOG-STANDARD BUSINESS

been entrusted with a business to run and he's drinking. And right before he's got an interview. Plus, the leather jacket.'

'What's wrong with the leather jacket?'

'It's casual,' I explained. 'It's what you wear in your free time. The guy shows up for a job interview wearing what he wears out with his mates on a Saturday night. That means he expects us to take him as we find him, and he'll expect our customers to do that as well. I'd rather an old-fashioned, threadbare suit if that's all he can get his hands on than an expensive leather jacket. It means he knows what formal means and he's willing to dress the part.'

On appearance, I've always stuck to my guns. No visible piercings, tattoos, no mohicans, ponytails or wacky dye-jobs. 'Thou shalt be presentable,' the Pimlico Bible says. In many ways this has been the toughest position to defend. People see it as their right to look how they want, and in their own time it may be. But everything is a uniform; everything sends a message. What does a nose ring, or pink hair, say to a customer? It says, 'I'm a rebel, I don't conform.' And what sort of stranger do you want in your house when you're stressed and vulnerable, with water flooding the kitchen and the kids asleep upstairs? Do you want edgy-looking rebels, or nice, presentable conformists? Do you want people in your home who are defensive about their comfort zones, or willing to step respectfully into yours?

At one point in the nineties ponytails came into fashion big time and I had to face down a full-scale rebellion. I don't know how it slipped my attention that everybody's hair was getting longer, but it did, and then one day all these guys came in with ponytails – about a third of the workforce, or so it seemed.

FINE TUNING

'Right,' I said, 'haircuts, the lot of you!' And they kicked up a right fuss, saying they wouldn't, saying I couldn't make them, that it was against their human rights, and all the rest of it.

But I didn't back down; I gave them a deadline. I said, anyone who comes in with a ponytail on that day gets the sack. And on the day I came in early and waited. It was like the shootout at the O.K. Corral. I was nervous – it would be a major disruption to let so many plumbers go, but I had to take a stand. Thankfully, when they started arriving everyone was neat and tidy. I had to sack only one person that day. One guy put me to the test. He said he'd sue me. 'Go ahead,' I said. I never heard from him again.

It was times like these that made me glad I had Lynda, and Scott by then as well, who'd come on as operations manager, fighting my corner.

Anyway, I would tell the HR people to spell it out clearly on the phone when setting up interviews, both to save time and to avoid nasty scenes. One day, however, the system failed and in walked this bloke with a ponytail – all the way down his back. It was like having a horse in the office!

'Well, he's here,' I thought, might as well have a chat. And it was pleasant to talk to him. He had the right sort of attitude, the right qualifications, and good experience.

'Thing is,' I said, 'if I give you this job, the ponytail would have to go.'

He looked at me, dead serious. 'Mr Mullins,' he said, 'I haven't had my hair cut in twenty years, but if you give me this job, it goes.'

I'd heard that before. People would promise to get their hair cut or take out the earring and then show up with hair still long or the earring still in, hoping I'd let it pass, which normally led to a row and them getting the sack. But I liked this guy and against my better judgement I called him up to offer him the job the next day. His wife answered. 'He's not here, Mr Mullins,' she said. 'He's down the barber's, having his hair cut. I don't know what you said to him – I've been on at him for years to get that thing cut off!'

I was so tempted to wind him up and say he hadn't got the job.

My goal is to revolutionise the plumbing industry. Back when I worked with Bill Ellis there was a lot of snobbishness about tradesmen. When he knocked on someone's door, if it was a new customer, as often as not the person would pull a face and tell us to go round the back. But if it was a repeat customer they'd throw the door open, smile and offer him a cup of tea. That taught me that if you show respect, you'll get respect back. Back in the 1990s it was unheard of for a tradesman to put a dust sheet down, or bring a Hoover, or pack up the debris – all of it – and take it away. It's still very rare – it's what sets us apart.

And it still works for us. How do I know? With every new customer we get, and keep.

CHARLIE ON... DOES THE SUIT MAKE THE MAN?

When author Mark Twain wrote 'clothes make the man', I'm sure he wasn't referring to the chosen attire of some young people turning up at Pimlico Plumbers for job interviews in 2014.

In fact, Mr Twain would have choked on his cornflakes, or whatever was the breakfast of choice in 19th-century Missouri, if he'd seen the state of one candidate who arrived at our depot last week in the hope of landing a job with us.

I know we've been experiencing this year's first wave of hot weather, but that is no excuse to arrive at an interview dressed as if you're going to the beach!

This guy turned up in shorts and T-shirt for a formal interview and was turned away before he even had the chance to sign in at reception due to not presenting himself appropriately for our company.

First impressions are everything, especially when applying for a job. But, in truth, in business, they are everything 24/7! Here at Pimlico, we pride ourselves on presentation with all our engineers being very smart and clean at all times.

From their immaculate uniforms to our gleaming fleet of vans, how our people present themselves to customers and the general public is a vital part of our business proposition, helps build our brand and demonstrates our commitment to quality.

Of course, having staff wear branded uniforms isn't the only way to ensure they are well presented and make a good impression. Before the recession a great deal of workplaces were influenced by the casual approach to business attire of West Coast of America Internet companies, which led to offices becoming awash with men in chinos and polo shirts.

However, just when some thought the only place we'd see the great British tie was in a fashion exhibit at the Victoria & Albert Museum, the economy crashed and there was a shift: the business suit made a comeback.

Why? Simply, because business people realised they had to work harder to win new work in a retracted economy and making the right first impression had to be the first weapon pulled out of their arsenal if they were in with a chance of landing a sale.

The same is true in the competitive jobs market. While there are more people in employment than ever before, there are still plenty of people out of work, particularly the under-25s, and jobseekers need to make sure they don't put themselves at a disadvantage against other candidates. That includes, of course, dressing correctly for interviews.

But before anyone tries to tar me with the brush of being a generalist, I know that all young people aren't scruffy and many understand how to dress for the workplace.

In fact, I'd like to end on quite a heart-warming story. We've been working with an organisation called Kids Company, which helps young people from vulnerable and disadvantaged backgrounds, to provide work experience places.

We currently have one of their lads, Adan, with us for two weeks' work experience. He's keen as mustard and the complete opposite to the beach-bum who turned up for interview.

We've kitted him out with a Pimlico uniform while he helps out one of

FINE TUNING

our plumbers, but that hasn't stopped him turning up each morning in a shirt and tie.

Undoubtedly, plenty of young people looking for a job as well as overly-casual workplaces could learn a lesson or two from Adan — his first impression definitely made an impression on me!

Originally published on 27 May 2014 on realbusiness.co.uk — The UK's leading title for high-growth businesses and entrepreneurial SMEs.

CHAPTER THIRTEEN

SAIL STREET

Whenever I do the thing that isn't normal, the thing that nobody else does, it comes out better.

In the latter half of the nineties the recession became a distant memory. Pimlico Plumbers was doing what it ought to be doing, and business was great. In 1992 I'd been terrified of losing the house in Sidcup, but in 1997 I sold it and moved to an even bigger one in Bickley, which I'd bought for £350,000. By 2000 we had so many people and so many vehicles that we were getting cramped in Lambeth Road.

It was a nice problem to have, but still a problem. For a while I considered getting another railway arch to make more room for the vehicles but then realised it was doing my head in, having the operation spread out over two locations. Three would be worse. It didn't seem like the way to run a proper business, all ramshackle, with bits patched on here and there.

BOG-STANDARD BUSINESS

Then I got a call from an estate agent I knew, telling me I should come and look at a building round the corner, some sort of old warehouse. So I did. It was huge — more than 30,000 square feet over two floors. There were sitting tenants: a coach company, a debt collection agency, and on the second floor this guy who bought and sold vintage cars. He rented them out, too — he had a lift for the cars and everything. The estate agent said they wanted to lease the place for £200,000 a year.

When he told me that I thought he was mad. Either that, or he reckoned I was a fool. I wanted to run off down the street. Two hundred grand a year, just to lease the place? How would I fill it up? What would I do with the tenants? It seemed crazy.

But I couldn't stop thinking about it. It was steeped in local history. Built in the 1930s, it had been a Sainsbury's warehouse, and a British Telecom warehouse. I found out that parts of the 1949 film *Passport to Pimlico* had been filmed there, as well. At night I used to go and walk around the block, looking at it.

Finally, I took my marketing manager Tony to see it.

'Buy it,' he told me.

'What?' I said. 'Don't tell me you're mad and all!'

'Just buy it. This is the future of Pimlico Plumbers.'

Part of me really didn't want to hear this, and part of me really did. No plumbing company in their right mind would take on a building this size. It wasn't what plumbing firms did! And it was exactly that, plus all the ballooning ideas I had about what I could do with the place, that made the notion so terrifying. I was scared in the way people who are terrified of heights are afraid — not of falling, but of jumping in the first place.

Yes, said the estate agent when I rang him up, the owners

SAIL STREET

would be willing to sell, but it would be a sealed-bid auction with a guide price of £1.8 million. There was a lot of interest from developers in such a big place in a prime location. So I bid £1.81 million, and won.

I was in a daze when I looked round 1 Sail Street for the first time as the new owner. At first I thought, three tenants, great, handy to have the rent coming in. But fairly quickly my thinking changed. My ambition for the business now had a physical shape — the building — and I wanted to fill it.

The place had been haphazardly managed for some time. I was very businesslike, and said the rents would be going up. Two of the tenants, the coaching company and the debt collection agency, said no thanks and started preparing to leave. The third tenant, the guy with the vintage cars whom I'll call Toby, was happy where he was.

A lot of people were warning me, 'Watch him, Charlie! He's shrewd, he'll have you over.' And I'll admit, that put the wind up me. He was a posh, public-school type, very smooth and charming. I couldn't work out how he made his money, shifting a few cars here, renting a few there. What tricks could he have up his sleeve?

He didn't have any tenancy agreement. This was clever — it's much harder to get rid of someone if nothing's written down. He would bung the former owners some cash now and again, and the rent was a pittance compared to the market rate — £500 a month. So I proceeded cautiously. I drew up an agreement setting the rent at £500 a month, but with increases every year until it was up to £1,000 a month in three years, which was more like what it should be.

'Oh, that's smashing, Charlie,' he said, signing right away. 'I couldn't agree more. I'm very happy here and don't ever want to leave.'

'Oh yes,' I said, 'and I don't want you to leave either.'

What was he playing at?

I had other things to worry about at the time because we were racing to get the bottom floor kitted out for the vans and the workshop, but I kept thinking about what I'd do if he started to put the moves on me. What were his weak spots? The elevator, I thought. I'd take the fuse out of the elevator so he couldn't get his cars in and out. He wouldn't last long then. That's how I was thinking: there was no way he or anyone else was going to have me over.

In the first year he paid the rent just fine, but in the second year, when it went up, he started to fall behind. 'Here we go,' I thought. There was a woman in accounts who did credit control and she was so fierce she made Rottweilers look like pussycats. I told her to get on to Toby. 'If we let him fall behind he'll have us over a barrel,' I thought.

'I say, Charlie,' he said to me one day, 'you couldn't have a word with that woman of yours, could you? She's frightfully unpleasant.'

'Don't worry, Toby,' I said, 'it's nothing personal, she's just doing her job.'

Within a month he came to me and said, 'I'm awfully sorry, old chap, but I seem to be having trouble with the rent. I think it's best for all concerned if I go.'

'Are you sure, Toby?' I said, making out this was terrible. 'I'd hate to lose you. I'd love to drop the rent, but I don't think I can.'

'I wouldn't dream of asking such a thing,' he said. 'You've been awfully good. No, I think it's time for a fresh start.'

And that was that.

The thing with Toby showed the power of mental barriers. People made him out to be this cunning foe, the engineer of my downfall. Why did that idea have so much power in people's minds, mine included? I think it's because I was upsetting the apple cart. He was an upper-class gent and I was a plumber from the Elephant & Castle. What was the world coming to if guys like me could unseat guys like him, take over so much prime London real estate and run a successful business out of it? Somewhere deep down it made people nervous, maybe even me.

But I was determined and in the end Toby was just a guy, and his crummy little business stood in the way of my growing business. In the face of normal tenant-landlord procedures he just went away. There were no plagues or wildfires, the world didn't blow up: it was the new normal.

How much do mental barriers like these keep people down?

It took two years altogether to refit the place and get the whole operation moved in, and it transformed the business. This was Pimlico Plumbers growing up, going to big school. With the space there to fill, we grew to fill it. We had space for our own panel and body shop for the vans, so I put one in. We also had space for the best welfare facilities – canteen, showers, changing rooms and the like – that London tradespeople have ever seen, so I put them in. And we had space for a state-of-the-art call centre – I hired our first full-time IT manager in 2002 – so in one went. We had space for a proper cafeteria for everybody, so why not get one?

BOG-STANDARD BUSINESS

No plumbing firm I'd ever heard of had stepped into such a big space — it just wasn't the done thing. But as I said, when I do the thing that isn't 'normal' it always comes out better. I've always thought, 'If everybody does what everybody else does, how come everybody's not a millionaire?'

Just a couple of years after I bought the building a developer offered me £3 million for it. A lot of people would have jumped at that. I could have pocketed a million and retired to Tenerife but I had bigger plans. Number 1 Sail Street is Pimlico Plumbers' destiny. Thanks to this building we started thinking like the sort of company we were becoming: the largest independent service company in London. If we go anywhere, it'll be up, with extra floors on top. There's nothing stopping us from being the biggest and best plumbing firm in Britain.

CHAPTER FOURTEEN

WHY THE STARS LIKE US

There is one whole wall in our reception area filled to the ceiling with signed photographs of celebrities, thank-you notes to Pimlico Plumbers for sorting them out. Movie stars, stage actors, pop stars, sports celebs, TV personalities – if you can name them, there's a good chance they're on our wall.

It was a bit of an accident, me becoming 'Plumber to the Stars'. Yes, I did choose the Pimlico area to work in because it was close to home and posh, and I later broadened that out to Chelsea, Knightsbridge, and other top-drawer areas, but I never set out specifically to build a celebrity clientele.

It started when it was still just me and the joiner working out of one room in the basement of Sussex Street. In 1985 I got a distress call to fix a loo at a fairly well-to-do address. It had once been a grand house, but was divided into flats now, and a little run-down.

BOG-STANDARD BUSINESS

I went in and knocked on the flat door and this tall, very beautiful woman with startled, hazel eyes opened it.

'Hello, I'm Charlie, the plumber,' I said. 'You called about the loo?'

'Oh, yeah, hi, I'm Marie,' she said in an American accent. 'Come in, it's this way.'

She led me down the corridor.

'I don't know what's wrong with the damn thing,' she said. 'Every time I flush it makes this noise like an airplane taking off or something. Feels like it's going to bring the house down.'

She was dressed for relaxing around the house – jeans, loose blouse, no make-up – but I could tell she was somebody because she had this tremendous poise, even though she was flustered.

It took ten minutes to change the old ballcock.

'Okay, that should work now,' I called out.

'You're done?' she said, coming out of the kitchen. 'Already?'

I explained that it was just a knackered fill-valve and flushed, to show her. The loo behaved itself and did its job quietly.

'Wow,' she said, 'that's amazing!'

Then she got suspicious. 'So how much is that going to be?'

I said some ridiculously low figure, barely enough to cover the part and the petrol, and she gave me this brilliant smile.

'Wow,' she said, 'that's amazing too!'

She really was a stunner, but I kept my head. I was sure I'd seen her before but something told me to keep quiet. I didn't want her to know that I didn't know who she was, in case she was insulted. It was better, I thought, to act as if I knew exactly but was too polite to make a big deal about it.

I told her I'd send her the bill, and left. It wasn't until I got back to the office and called Lynda that I learned who Marie Helvin was: the famous model, recently divorced from the fashion photographer, David Bailey. Soon she would become a familiar face on television. I was well chuffed, and for weeks I went around telling everybody: 'I fixed Marie Helvin's loo!'

I was even more nervous when Joanna Lumley called. She was a proper TV star by then – I'd seen her on *The New Avengers* and *Sapphire & Steel*. As I said, this was around 1985, a few years before *Absolutely Fabulous*. But she put me right at ease. She sounded terribly posh but was actually very kind and funny.

The next 'face' to call me up was Reginald Bosanquet, for years the anchorman on *News at Ten*.

For a long time I couldn't work out why I seemed to be getting more than my fair share of celebrity customers. But then somebody told me – I think it was Joanna Lumley on another house call – that I was a regular topic of conversation at dinner parties in Chelsea and Knightsbridge. I couldn't believe it, but I guess it made sense. Celebrities need plumbers just like anybody else, and not many of them are so rich that they have underlings to handle all the domestic stuff for them. Sometimes they just have to pick up a phone and sort things out. At the same time, the paths they've taken in life haven't usually required them to be very practical – most couldn't bang a nail. And like everyone else, when there's a problem with the plumbing they feel stressed, embarrassed and vulnerable, so when I show up and provide courteous, respectful, no-nonsense service, they're amazed. Some have big egos, but when there's a problem, they're putty in your hands.

BOG-STANDARD BUSINESS

We hit the jackpot with Michael Winner, the film producer and director who'd done the *DeathWish* films and loads of others. He took a real shine to Pimlico Plumbers and recommended us to all the actors and celebs he hobnobbed with – pure, marketing gold. And if you can get along with Michael Winner, you can get along with anybody. He was so exacting! If everything wasn't dead right, he'd certainly let you know.

I'll tell you what I mean.

Normally we pick the plumber to suit the customer, but on one occasion the plumber who was Michael's 'usual' wasn't available, so we sent someone else. It wasn't a big issue from our point of view because all our plumbers are good and can be trusted, but certain types of people get along better with certain types of people and that's part of the service we provide. Anyway, the appointment was for eight in the morning and this plumber made the mistake of arriving early – he knocked on Michael's door at 7.53 am.

Michael was furious. 'I told you eight o'clock!' he shouted, plus one or two other choice things. Then the plumber called me in high dudgeon, saying he wouldn't work for that so-and-so, that he'd been rude and abusive, that life was too short, and so on. I had to calm him down. I explained that Michael was used to getting his way on huge, expensive film sets, that that was his job, that maybe he was on the phone with some big star, that he was a really important customer and a nice guy, really, and that all he, the plumber, had to do was keep an eye on his watch and knock again at eight.

Reluctantly, our plumber agreed, and when he did knock again, at one second past the stroke of eight, the door flew open and Michael was all smiles. 'Lovely to see you!' he said. 'Thanks for coming on time! How about some toast and tea?'

A small number of our celebrity customers have been wankers, I admit. We had a couple of our guys working in the London home of one big-name Hollywood couple (who shall remain nameless), and they had their butler or manager, or whatever hurry up to them and say, 'Mr and Mrs X will be coming along the corridor shortly, and you are requested to look away when they do.'

But for the most part celebrities are just like you and me. I think we get so much repeat business from that section of society because we're good, for one thing, but also because we know how to behave. We don't jabber away or moon about, all goggle-eyed. That makes anybody nervous. The thing about proper comportment, and by that I mean being respectful and courteous – but just enough, not overdoing it – is that it's an international language. It puts anyone at ease, whether you're a superstar in a Chelsea mansion or a granny in a council flat.

Now it's time to tell you how I became something of a celebrity myself.

CHAPTER FIFTEEN

HOW I GOT FAMOUS

I am often asked to speak at business conferences and seminars about PR, and I'm always a little surprised by it. Now, I'm not being modest. I know I'm seen as a whiz at publicity, and even before I hired Max Clifford in 2007 — when his asking rate was £30,000 a month — I was the best-known plumber in London, and Britain, and maybe even the world.

There is no doubt now — I am the World's Best-Known Plumber. I or my company have featured in five major British TV documentaries. We've also been on TV in France and Sweden, and we just waved goodbye to a TV film crew from South Korea. Hardly a week goes by when I'm not being interviewed on TV or national radio about something, and my staff are as used to seeing cameramen and guys with big furry microphones around the place as they are the cleaners.

In print, there are twenty-three articles about me in the

BOG-STANDARD BUSINESS

Independent, thirty in the *Guardian*, thirty-three each in *The Times* and *The Sun*, thirty-six in the *Daily Mail*, and forty-one in the *Telegraph*. Around the world I've been in *The New York Times*, the *Washington Post*, *Le Figaro* and *Le Monde* in France, *Gulf Times* in Qatar and *The Age* in Australia. Whenever I walk down the street, whether in London, Dubai or Marbella, I always get someone coming up to me, saying, 'You're Charlie, aren't you?'

But here's the thing: what competition do I have for the title of World's Best-Known Plumber? Absolutely none. Zilch. Why? Because plumbers don't bother with PR. They'll advertise, and some will bother with industry awards (like I used to, but don't anymore), but plumbers and tradespeople in general wouldn't dream of trying to be *famous*, like I am. They think it's weird, or undignified, or beyond the reach of normal people, as if my fame was just a freaky bit of good luck.

And yet I keep getting invited to conferences and seminars where I'm asked, 'Charlie, how do you do it?' What surprises me is why it's such a mystery to them, because it ain't rocket science. It also surprises me that service companies of all types care so little about PR because, when it comes down to it, the success of Pimlico Plumbers is based on two things: one, we do something better than anybody else and two, a lot of people have heard about us, and more and more people are hearing about us all the time.

Here's what I say at conferences and seminars when I'm asked about PR: Get Recognised. It doesn't matter what you do or how you do it, just get recognised. If that means running down the street buck naked with your logo tattooed on your backside, fill your boots! Fortunately for me it never came to that, but then again, I'd try almost anything.

I've always had a nose for publicity. Getting Roger Bannister to check me out after my fall in the boxing ring was a move calculated to attract the press, for instance. Even then I saw publicity as a tool and wasn't afraid of using it. But it was Tony, the graphic designer who became my marketing director, who turned me on to its possibilities for the business. Recognition: anything that snags the fickle attention of the general public is worth doing, whether it's brightly-coloured uniforms, wacky stunts, or your logo on a coffee cup.

Do I have to explain why it's worth doing, whether you're a man with a van or a multinational corporation? Given how few small companies bother with it, maybe I do. It's not very complicated: if you're in people's minds when they have a problem, they'll call you. If you're not, they won't.

Once you have that mindset, the possibilities are endless. People are quirky, so there's no telling what'll snag their attention. All you have to do is keep trying things, however hare-brained.

For instance, in a million years I couldn't have guessed what a wild success my first go at it would turn out to be. Back in 1993, when we were still hanging on for dear life, the chance came up to buy a personalised number plate that said 'DRA1N'. They wanted £6,000 for it, and wouldn't budge on the price. Six thousand pounds for a number plate! Of course I knew it was crazy and yet I couldn't let go of the idea. I had sleepless nights over it but in the end I coughed up.

It was a hit. This was before it was common for businesses

BOG-STANDARD BUSINESS

to have personalised plates, and people loved it. So I just kept collecting them. We now have over 130, and I've paid as much as twenty-five grand for a single one. It's worth it, though. For one thing, it's an investment. People are always after me to sell. I could easily get a hundred grand for DRA1N now. Some while back I had a call from some people representing Roman Abramovich, the Russian billionaire who owns Chelsea Football Club. They wanted to know if I'd sell B0G 1 because, as it turns out, 'bog' is the Russian word for 'god'.

But I'd never sell because the plates are even more valuable as marketing tools. They have a huge following – why, I don't know. I guess it's a bit of humour out on the streets, the faceless vehicle registration system winking at you. Every week people take pictures and Tweet them or put them on Facebook. They book our plumbers by the number plate – they'll ring up and say, 'I don't know the plumber's name but it was S1NKS.' Some people are number-plate spotters and mark them down in a book. 'I've had B1DET and I've had 701LET,' they say, 'what I'd really like next is S110WER!' One woman wrote us a letter to say that she sees our vans with the funny number plates every day and it brings a smile to her face. She said our vans were part of London now, like Hackney carriages. I was very proud to hear that.

Another thing that got us recognition was our model vans. How many other plumbing companies do you know that do a roaring trade in their own model vans? No, me neither. The idea, of course, came from the British haulage company Eddie Stobart, which is easily far more famous for its miniature vehicles than it is for hauling stuff. Before I had a head for publicity, and was just a successful plumber, I remember thinking about Eddie

HOW I GOT FAMOUS

Stobart model lorries and saying to myself, 'Wow, imagine if we were ever famous enough to do that' (which is the wrong way to think of it). Then, about six years ago, we started doing it. And they absolutely took off.

I insisted from the start that we had to do it right. We were a great company doing great things, so the vans had to be great. They were going to be highly detailed and highly accurate, right down to the number plate. Tony Davidson designed them, and it's no simple thing, making a full-size van recognisable in miniature. When we launched them they got great reviews in collectors' magazines (I had no idea there were magazines about model vehicles) and they were appearing on eBay for five times the asking price, which is £10. Because of the number plate, each batch is a limited edition, and they sell, literally, all over the world. We order a batch of 3,000 every few months.

I don't know how much money we make on our model vans. There was a rumour going around that, at one time, Eddie Stobart made more from its model lorries than hauling stuff. I have no idea if that's true, but it certainly isn't true for us. What you can't price, however, is the buzz they create and the recognition we get from them.

For a while in the 1990s we went in for plumbing industry awards and of course we won quite a few. It was an important part of our PR journey because it gave us our first taste of trade and local press coverage but we gave it up in the end because it's a mug's game. For one thing, these awards are usually rigged. All too often I've seen companies win not because they're the best but because it's their turn, or because the organisers want their sponsorship, or for some other dubious reason. Maybe

I'm arrogant, but I believe we could win hands down any award we went in for, and I saw us passed over one too many times. It reminded me of boxing.

The other thing about plumbing industry awards is, the general public doesn't give a stuff about them. I mean, it does add a bit of kudos to be able to say you've won this award or that award – we won National Installer of the Year twice – but does the public actually care? No. Plumbing industry awards are an industry in their own right. They are two-a-penny and any buzz they create stays inside the industry. Meanwhile, there are things that really do capture the public's imagination. That's the big game, and that's what you need to go for.

So here's my philosophy on publicity:

1. It's good.
2. You can get it.
3. If you're in business offering a service to the general public, you are letting your business down if you don't do everything in your power to get it.
4. Publicity doesn't come by fluke or accident – it's something you need to work on, all the time.

What you need is a publicity head, and I'll explain what I mean by that.

Doing the Croydon Fun Run in a Mr Blobby suit to raise money for Children in Need is great – it will get you in the local paper. (Reporters on local papers are so bored they'll cover anything.) It will tickle the fancy of some of the general public for a couple of minutes. It's great – 100 per cent better than nothing. But that was last week. What are you going to do to tickle the fancy of another bit of the general public this week? And next week? And the week after? What are the next ten things you're going to do to get yourself noticed and talked about? That's what I mean by a publicity head.

Also, at some point you need to start thinking about how to get more return on your publicity effort. Running around in a Mr Blobby suit for a mention in the local paper is high-effort, low-return. It's a start, but you'll burn out pretty quick if you don't hit on some better ideas.

We put in the publicity legwork throughout the 1990s and early 2000s, getting trade and local press coverage with the awards and charity stuff. By then we were pretty well-known for our celebrity clientele, too but we got our first big break with the BBC TV documentary, *Posh Plumbers*, in June 2004. The idea was to film bankers and other middle-class graduates giving up their ninety-grand-a-year salaries to train as plumbers, and when the producers looked around for a suitable plumbing outfit to use, it didn't take them long to find us.

Of course I agreed to do the programme but then I panicked. I got it into my head that they had a hidden agenda; that they'd make us look bad somehow. Isn't that what TV programmes did? Looking back, it seems pretty paranoid now, but it was all so new to me. I had all but decided to pull the plug on it when I called one of the producers, a woman, and explained my concerns. She was very good.

'Charlie, the point of the show is posh plumbers, not whether Pimlico is a good company or not,' she explained. 'And anyway, if you're a good company like you say you are, how can we make you look bad?'

I was fine after that – she'd put me at ease. The show went

BOG-STANDARD BUSINESS

ahead and was broadcast at peak viewing time, nine in the evening on a Tuesday, and got 3.5 million viewers. There were also all the previews in the national press and on TV before, and the reviews afterwards. We were famous! And I was pleased with how it came out. We looked... well, we looked like us.

After apprenticing with us one of the stars of the show, Matthew Brumwell, who'd left his high-powered job with an investment bank in the City of London, set up his own plumbing business in Cheltenham. He seems to be doing well, and the stated mission of his business, Cotswold Plumbers, is to erode the image of the plumber 'turning up late, unkempt and without the right tools'. Sounds like something of Pimlico Plumbers rubbed off! Good luck to him.

From an advertising point of view, *Posh Plumbers* was priceless. Primetime TV ads cost something like a hundred grand a minute, so sixty minutes was worth six million quid!

But if I thought I could sit back and bask in the limelight I had another thing coming. Three-and-a-half million viewers is fantastic, but how many of them lived in the areas of London we served? And of those who did, how many needed plumbing services right there and then? Calls certainly picked up, including from places like Scotland and Cornwall, which was not much use to us. Also, the soufflé of fame falls very fast. People hear about you and right away they start to forget. Publicity is like a tyre with a slow leak – it needs constant pumping up. You have to keep feeding the beast.

After the coverage died down I started to get impatient, wondering what to do next. Then on a hunch I did something that may well turn out to be the single most effective PR move of my life – I hired Buster Martin.

CHARLIE ON...
EFFECTIVE MARKETING

People are always interested to know what my secret to success is, and what makes a Pimlico plumber in some way different from any other plumber. The answer is hard to put into words, since it's like asking Warren Buffett why he was prepared to shell out **£18 billion to buy Heinz**, when other bean canners were available, and for a damn sight less money.

We all know that '**Beanz Meanz Heinz**' was a result of a ridiculously successful advertising campaign dreamed up, as legend has it, by Maurice Drake in 1967 over a couple of pints in a Victoria pub. But even back then – 46 years ago – this slogan was only stating what people already knew: when someone said baked beans, people automatically thought Heinz.

And while it would sound pretty cool for me to say now that when I started out in the plumbing business my plan was to become the Heinz baked beans of the plumbing world, it would be a barefaced lie. And I suspect it was no more true for **Henry Heinz** when he founded his food company in 1869. But he, like me, I'm sure, set out to run the best operation he knew how, which in 1901 led to the introduction of the best, most popular beans ever baked, tinned and smothered in tomato sauce.

To begin with, I did my apprenticeship and learned my trade as a plumber and gas engineer, during which time I learned a lot about what people thought about plumbers, even if at the time I didn't realise it. But when you tell people you are training to be a plumber and their immediate response is some kind of reference to Dick Turpin, you are

BOG-STANDARD BUSINESS

bound to pick up the idea that your chosen profession isn't exactly held in high esteem. I wanted to tell people I wasn't like that – who wouldn't? And the best way to do that was to buck the stereotype, by doing a good, honest job.

It worked. People told their friends and neighbours about me, and things started to snowball. Back in those early days, I was working a lot in Pimlico, and people started calling me 'the Pimlico Plumber'. So when I had to come up with a name for my fledgling company in 1979, **Pimlico Plumbers** came into being. In a way, that was when my version of the famous Heinz baked bean tin was born; before that I was a decent plumber, but after I left a customer's home, I was just that funny cockney bloke with a bag of tools. Now I had a banner, the catchy name 'Pimlico Plumbers', to stick in people's minds, long after they had forgotten my name.

Since then, increasingly the marketing of Pimlico Plumbers has become all about cramming all the positive aspects of our unique service into the brand, like so many beans in a tin. The thought is that whenever someone hears the name, sees our logo, or spots a Pimlico Plumbers' van that all the good things that we have packed into our name will pop up like a jack-in-the-box.

People ask why I've spent so much on more than a hundred number plates such as BOG 1, DR4IN and W4TER, and the answer is they give people a fun reason to look out for and notice our vans. More recently, we have had model versions (also with different number plates) made, which are also hugely popular with our customers.

Basically, we have established what we stand for over thirty-five years in business and now what we are doing (apart from doing great plumbing, building, roofing and electrical work, etc.) is making sure

126

people don't forget the brand that stands for a quality job, done by a fully trained, polite tradesman, in a clean corporate uniform, driving an instantly recognisable van.

It's taken a lot of years of hard work to get where we are today, and having created a premium reputation within London, it takes a lot of paying attention to detail to maintain it. But I would like to think that whenever someone needs a plumber, they recall seeing S1NK drive past the other day, and think: 'I've heard a lot of good things about Pimlico Plumbers, I'm going to give them a call!

First published by www.theguardian.com on 6 March 2013

CHAPTER SIXTEEN

BUSTER

In 2004 we advertised for a general dogsbody to help out around the place, mostly valeting the vans. We got lots of applicants, but one in particular stood out.

I'd never met anyone like Pierre Jean 'Buster' Martin before. His story was remarkable. He said he'd been born in the Basque region of France when his mother, a poor woman, had fallen pregnant to some dude from a local well-to-do family. They smuggled her to Cornwall to avoid the fallout and little Pierre was put in an orphanage near Bodmin, where he got his nickname for belting a priest when he was three. When he was ten they kicked him out – because he was too big and hungry, he said – and he went to London and scratched a living doing errands for the traders in the Brixton market (which I could certainly relate to!). He joined the British Army when he was fourteen, and wound up as a physical training instructor. Having

BOG-STANDARD BUSINESS

fought in the Second World War, he left the Army in 1955 with the rank of regimental sergeant major. He was married for thirty-five years and had seventeen kids, but his wife was long dead and the kids were scattered all over the world.

After the Army he'd worked in various jobs but ended up back at Brixton market, organising the stalls. There he worked for years and years. He'd only retired three months before he came to us.

'What do you want to work here for?' I said. 'You should be putting your feet up.'

He fixed me with his slightly crazy eyes. 'I am going out of my mind with boredom!' he told me.

He was ninety-seven.

'What the hell,' I thought, and hired him on the spot.

Now, did I hire Buster purely for his publicity potential? I wouldn't put it quite like that. I liked him and admired his attitude. I thought his wisdom and experience would be useful, and I definitely did not like the idea of someone like him, with all his marbles intact and plenty to offer, just wasting away in the little flat he had round the corner. But I'd be lying if I wasn't also thinking, 'Hey, this'll give us something interesting to talk about!' I don't know what I expected, exactly. All I knew was, the tyre of publicity had to be pumped.

But the flood of media interest was slow in materialising. Buster got some coverage in the local press the following year, but that was it. And this is where I learned my next crucial lesson about publicity: just as in plumbing, if you want the job done right, you need to call in the professionals.

In late 2005 we had some automatic security gates installed

on the premises by a specialist firm from up north. They'd done a good job, and when they were finished I got a call from a PR company asking if I'd mind if the MD of the gates manufacturer, a woman, came down for some photos.

No problem, I said.

So she came down and after the photo shoot we were chatting. She'd heard a little about Pimlico Plumbers and she asked, 'So who's your PR company?'

'What planet is she on?' I thought.

'PR company – what would I want a PR company for?'

She explained that hers was going to put these pictures and a feature article in a magazine up north and everybody would read about this job, and about me.

'Really?' I said.

'Yes, really. You can't do PR without a PR company. They know how to spin a story and how to get the media interested.'

'Well, f*** me,' I thought. 'It's so obvious!'

She left me a card for Recognition PR – that hit home, 'recognition' – and we've been with them ever since. (In 2007 I also hired the publicist Max Clifford, but a publicist like that does different things – and costs a lot more – which I'll explain a bit later.)

In June 2006, when Buster was ninety-nine, Recognition put out a press release about 'Britain's oldest employee' working for Pimlico Plumbers and being told by his bosses that he had to take a day off for his one hundredth birthday in September. Three days later *The Sun* ran the story with a big picture of Buster and the floodgates opened.

After that, Buster did clean vans when there were cameras

around, but his real job with Pimlico Plumbers was pretty much drinking beer, smoking fags, and telling stories – just being himself, in other words. He was a complete natural at that, and the media loved it. He turned one hundred and it was all over the news. We threw a big party for him at Chelsea FC's Stamford Bridge ground.

The following year, 2007, on the way home from the pub one night, he fended off an attack from three young hoodlums and spent the night in hospital with a bump on his head, and was all over the news for that. Liberal Democrat bigwig Ming Campbell publicly praised him as living proof that age shouldn't be a barrier to work, and that made the headlines, too. Also that year, the men's magazine *FHM* took him on as an 'agony uncle' columnist, and he was invited to join The Zimmers, an oldster rock band that shot to fame with their cover of 'My Generation' by The Who, which was produced by U2's producer Mike Hedges and used in a BBC documentary.

Somehow, in between appearing on TV, meeting celebrities and giving politicians a hard time, Buster took up running. Supervised by my son Samm, a Pimlico Plumber and a boxing coach, Buster ran the 10km Bupa Great Capital Run in 2007 and the Roding Valley Half Marathon in 2008. The media couldn't get enough of this drinking, smoking, red-meat-eating one-hundred-and-one-year-old geezer running long-distance races – 'shuffling' might be a better word – and Buster couldn't either, apparently, because he then entered himself into the 2008 Flora London Marathon!

It was a shame that the marathon was in April because, had it been on or after the first of September, he'd have been one

BUSTER

hundred-and-two. Never mind, at one hundred-and-one he'd still be the oldest recorded marathon participant in the world, and we duly approached the *Guinness World Records* people to let them know Buster would be expecting his rightful entry. We also got the famous peace activist Harmander Singh on board. Harmander was a famous marathon runner and trainer for marathon-running old people. What clean-living Harmander thought of our loveable reprobate Buster I'm not sure, but the two seemed to get along very well.

A few days before the race Buster – a confirmed gambler to boot – also slipped into the bookies, William Hill, and placed two bets, one on himself, the oldest marathon runner ever, finishing the race in under twelve hours, and another on him crossing the line before midnight. The odds were long – 100/1 and 33/1 respectively – and the payout, should he win, he pledged to a sick children's charity, the Rhys Daniels Trust.

On the day, Buster finished the race in 9 hours and 59 minutes, fuelled by five pints of beer along the way, each accompanied by a cigarette! We were over the moon, but when we went to the Guinness people they refused to recognise his achievement unless he produced a birth certificate. Buster didn't have one. We got him a passport and a naturalisation certificate, but that wasn't enough, they said. I even hired a private investigator to look into it, but he came up blank. Tracing an illegitimate child born in France and smuggled to Britain to an orphanage that didn't exist anymore, all before the First World War, would have defeated even Sherlock Holmes.

This, too, was all over the news.

Then we received another blow when William Hill jumped

on the bandwagon, saying that if Buster couldn't prove his age there was no way they would pay out his winnings – a total of £13,300. All through that spring and summer we fought a campaign in the press to get William Hill to do the right thing and give the sick kids their cash. We even got in top City lawyers, Mishcon de Reya, to threaten legal action. Finally, in August the bookie decided to give the money to the Rhys Daniels Trust as a donation, not as a payout. Honour was satisfied on both sides, and the Trust got its much-needed money. Even before that was settled, Buster was running again, doing the 2008 Bupa Great Capital Run in Hyde Park for Capital Radio's 'Help a London Child Appeal'.

I can't count the number of times Buster was on TV. Crews from all over the world came to capture him on film. But my favourite was the film called *How to Live Forever* (2009) by the film-maker Mark Wexler. He spent three years going around talking to all kinds of people over one hundred years old, some of them pretty eccentric, including celebrities like Phyllis Diller and the writer Ray Bradbury. It was all about the secrets of immortality. He talked to Buster, too, who solemnly explained his secrets, including smoking, drinking only beer and coffee (not water, ever, even during marathons), avoiding fish and sticking to red meat, and so on. I think it was Buster at his best, like he was being himself for once and not playing the cheeky old chappy to the camera. He became the poster boy for that film, and it was a hit around the world, premiering at the Hamptons International Film Festival in 2009 and screening at the Palm Springs International Film Festival and the Gasparilla International Film Festival in 2011.

BUSTER

But of course even Buster wasn't immortal. In 2010, at the age of one hundred and three, he was riding his bike when he went into the side of a bus. Nothing was broken, thank goodness, but he was battered and bruised. He recovered in due course but he wasn't quite his old self anymore, never as ready to break into a trot as he had once been. A year later, on 12 April 2011, just four months short of his one hundred-and-fifth birthday, he died at home in the night.

Once again, Buster was all over the news, though this time it was his obituary. For his funeral we lined up every available Pimlico Plumbers van — the procession was about a half-mile long — and drove from our headquarters to the crematorium in Camberwell. There were whispers of criticism about me milking his death for publicity, but that's totally wrong. We were the only family he had. He was known and loved all over London and the world. He'd done so much for the business, for us all, and I felt that it was perfectly fitting.

A lot of people puzzled over Buster. Once in a while journalists would have a go at picking holes in his story but nobody wrote him off as a nutter because it was clear that he wasn't one: he was a smart guy. His life-story was colourful and, whether deliberately or not, completely impossible to verify. Sometimes the more pints you bought him, the more outlandish the stories got, like the one about him 'borrowing' a Spitfire during the war. But his memory was crystal clear, both short-term and long-term. Buster never forgot a detail, and he never forgot a face or who he'd told a story to. He read well, and was a crossword fanatic to the day he died.

What I think happened was this: he was offered a role and he was happy to take it, but on his own terms. With the things he said about drinking, eating meat and smoking, you got the sense that it was him taking the piss out of the rest of us.

'I'm not like you people... normal,' he told Mark Wexler in the film. Then he chuckled, puffed his fag, and looked away for a minute before adding, 'And I mean that.'

Did I exploit Buster? The value of the publicity he brought is probably incalculable, although, as I've said elsewhere, our sales increased by around 36 per cent between the time of Buster joining us and his death. For his part, in his declining years he had a comfortable living, companionship, a purpose in life, and a first-class whale of a time. If there was any exploiting going on, I think it's fair to say it went both ways.

CHARLIE ON... AGEISM

We often hear about how the over-60s get a raw deal in the job market and the challenges that young people face getting onto the first rung of the career ladder.

But, now, apparently, it's also the over-40s that are finding it tough to get on in the workplace.

A survey by a law firm seemingly peddling its workplace discrimination services from the back end of last week revealed that a fifth of over-40s feel their careers are stalling because of their age.

Of the 2,000 people questioned, one in seven felt they were more likely to be overlooked for promotion than younger colleagues, while another 23 per cent felt they had hit a 'brick wall' in their current job.

While I'm not quite sure of exactly how representative of the workforce this kind of survey really is, it does suggest that pretty much everyone has a beef with employers. Which is no surprise, as it seems that apparently everyone's either too young, too old or too middle aged!

Young people say they are overlooked because they don't have any experience and without a job they can't get experience. The over-60s claim ageist employers want youth over experience, which now seems to be a view shared by those in the 40-plus bracket.

Surely, they can't all be right? But the fact is, to a degree, they all are. We all know, and are told by law, that no one should be discriminated against at work for their age or anything else. Unfortunately, since the dawn of employment, that hasn't been the case.

And even in today's increasingly litigious society it appears that in some, of course not all, businesses discrimination is alive and well.

Of course, that doesn't happen at Pimlico Plumbers, but you'd expect me to say that. But the proof, as they say, is in the pudding.

I am a consistent campaigner for giving young people the opportunities to gain experience by creating as many apprenticeships as I can. Also, 25 per cent of my workforce is aged over fifty-five as older workers are a resource not to be ignored.

Jobs at Pimlico are filled by people that have the skills and aptitude to fit the role. It is a position that every business, and most do, stick to. That means that not everyone is successful in getting a job or promotion, but that's life in the big, bad real world.

Perhaps we are just so used to hearing this sort of 'woe is me' language in surveys and social media posts that it has become seen as the truth about how people are treated in the workplace, particularly

BOG-STANDARD BUSINESS

from those already in work who are searching for the nirvana of 'total job satisfaction'.

But instead of complaining about the firms that won't advance their careers, they should put their efforts into making themselves more attractive to employers, which might give them the edge the next time they go for an interview or promotion.

Originally published on 24 February 2014 on realbusiness.co.uk – The UK's leading title for high-growth businesses and entrepreneurial SMEs.

CHAPTER SEVENTEEN

CHARLIE'S ANGELS

Buster showed me how hungry the media were for stories about social issues – in his case, age discrimination and old people throwing off the shackles. So while the Buster media storm was still breaking, I took a punt on another idea – female plumbers.

In late 2006 we started recruiting female apprentices and taking on female plumbers. It's not that we discriminated before, but in such a male dominated industry they were not exactly beating a path to our door. What changed was, we went out looking for suitable candidates. We were probably one of the first companies besides the Gas Board to do this. In January 2007 we kicked things off by getting three female apprentices to visit Norbury College in Surrey, a business and enterprise college for girls, glammed up to the nines and chauffeured in our best van.

I loved our lady plumbers to bits at first. And the public loved them too. I'd have women ringing up for them, old people, people that didn't like men, and people that live on their own. From a PR point of view, though, it was a bit disappointing. *The Sun* picked up on it and profiled our apprentice, Natalie Gowers, later in 2007. A couple of TV channels did news features, and the *Daily Mail* profiled our plumber, Jane Graham, in 2010, because she was taking home £80,000 a year. All good stuff, but not a patch on the publicity Buster got. All he had to do was wiggle his little finger to get the column inches, or so it seemed. I realise now that what really grabs the public's attention is people stories, not issue stories.

Nowadays we don't have that many female plumbers. The ones we do have are absolutely great, but I went a little mad with it and in practical terms it was a nightmare. For one thing, some of them wouldn't do certain types of job, like when there was an awful lot of stink and mess involved. Some didn't want to go to a flat full of blokes late at night, and who could blame them?

For another thing, it was causing issues in the workplace. In fact, it set the cat amongst the pigeons, sexually. Some of the men weren't prepared to work on the same job as a woman because they didn't want their wives to know. Other blokes were saying, 'I can't work with her, she's gorgeous! I won't be working, will I? I'll be chatting her up all day!'

Then there were complaints from the women about this fellow or that one getting too close. I'd say to the plumber, 'Look, you've got to behave yourself.' And the guy would come back with, 'I only squeezed past her in the cupboard!' And I'd be thinking, 'Yeah, I bet you did.'

CHARLIE'S ANGELS

One female plumber put in a complaint about sexual harassment, that this guy had grabbed her. When I confronted him, he said to me, 'No, Charlie, what we used to do was, we'd play a little game: she'd hide somewhere and I'd hide somewhere and we'd scare each other. So this particular time I was hiding in a cupboard and she came in and I just grabbed her to freak her out.'

For mercy's sake, they're playing hide and seek!

It was all becoming uncomfortable.

I still think women plumbers are great. I could probably start up a women-only plumbing company tomorrow and be a millionaire overnight with it. But from a practical point of view they were a bit of a disaster for Pimlico Plumbers.

CHAPTER EIGHTEEN

TV AND ME

People often ask me, 'Charlie, how do you get on TV so much?' and the answer is, I'm not entirely sure. Sometimes I wonder if there aren't more cameras in the world than there are half-decent things to point them at. For instance, there was a crew in here the other day doing a programme about our *canteen*!

But I suspect it has something to do with us being very media-friendly. After freaking out over *Posh Plumbers* in 2004, I'm not afraid anymore. We let it all hang out: we're outspoken and we've got nothing to hide.

Take the 2012 Channel 4 programme, *Show Me Your Money*, where the staff declared how much they earned, even me. It was voluntary, but lots of people agreed to do it, enough to make the idea work on TV. Everybody wrote their names, job titles and salaries on a Post-it note and stuck it up on a board. All on camera. I kicked things off: a million quid. People were reluctant at first but eventually the board was covered in Post-its. It was pretty shocking. Some people earned loads more than anyone would have guessed, some loads less, and there were big discrepancies — some people were taking home less or more than others even though they were doing the exact same job. Me excluded, on one end you had my son Scott, the operations manager, on £120,000 a year while Tina, who slogged away full-time in the canteen, was on £14,500 per annum.

The fur started flying that day, I can tell you! There was a lot of bitterness and people demanding pay rises. I was made out to be a villain because of what I paid myself — even though I'd started the company and made it what it was through years of toil and sweat, worry and risk.

My first reaction was to dig in. No doubt, we had a problem. There was no transparency in our pay structure. In fact, there was no pay structure at all. When it came to setting salaries and responding to requests for pay rises, it seemed we'd just made random judgement calls. But I wasn't prepared to cave in to every demand so I threw the ball back into the employees' court. 'You sort it out,' I said. I told them to get a committee together and see if they could persuade the bigger earners to donate some of their salaries to the smaller ones. 'Whatever they donated, I'll match,' I said. Some reviewers called this a cop-out on my part, and maybe it was. Divide and rule, you might say. But I don't respond well to demands and also I didn't know where to start in fixing things and thought it would be useful to see what they came up with.

What they came up with made great TV. There were arguments and refusals, and one selfless donation. When people didn't seem to be budging anymore the committee got them to swap jobs to gain a better appreciation of what others did. Karl, my PR manager, did a shift in the canteen and one of the top plumbers spent an afternoon manning the phones in the call centre. Both situations were hilarious! The plumber, who could find his way round a state-of-the-art boiler blindfolded, just could not get the hang of the switchboard and spent the afternoon getting earfuls from irate callers. As a PR professional, Karl is used to dealing with just about anything. A big part of his job is responding to unpleasant surprises, like the bad headline we didn't see coming, or being put on the spot by some reporter. I honestly don't think I've ever seen him in a flap — until that day, that is. There is not much that can prepare you for a mob of hungry tradesmen at seven in the morning with tight schedules to keep, except maybe experience. He had orders coming at him left, right and centre, people shouting for their bacon sandwich or their scrambled eggs, and after about fifteen minutes he was in meltdown. Tina was positively gloating. Afterwards he said the canteen staff dropped him in it by not doing the normal prep and basically abandoning him to his fate. We'll never know, I guess.

I did worry at one point about mass resignations because the tension was pretty high, but in the end things settled down. On camera there were a few more donations, some pay rises, and some of the more glaring discrepancies were sorted. Off camera, after the programme was aired, we got on with the more complicated task of making a logical, transparent pay structure. Some people who could not be appeased did leave, which was maybe no bad thing.

BOG-STANDARD BUSINESS

My point is, there aren't too many companies who would subject themselves to such an ordeal in the full glare of prime-time TV. In fact the producers picked us because we were the only ones willing to do so. They came to me saying, 'We want to talk about your pay structure for an hour on TV' and I was like, 'Darling, we'll talk about anything you like for an hour on TV!'

I wasn't quite so carefree with *The Secret Millionaire* in 2009. In fact I'd turned it down two years before. If you don't know the programme, it's when a millionaire pretends to be not rich and gets amongst some people who really need money and then springs it on them – hey, I'm a millionaire, have some cash! I'd turned it down before because I wanted to know what I would be doing and where my money was going. I didn't want to end up with a load of alcoholics, or in a prison or a rehab place. I had certain things I wanted my money to be doing, certain things I wanted my efforts to be associated with, and certain things I didn't. But they insisted, no, it didn't work like that, so nothing happened.

But then my son Scott and his family did *How The Other Half Live* and the producer of that show came to me and said, 'Charlie, I know how you feel, but if you could just trust us on this, I'll make sure you're not doing anything you really don't want to do.' He was good, so I said okay.

Mistake! I shouldn't have trusted him. Now I was thrown into the deep end of a situation I was totally unprepared for.

The deal was I would go to Warrington and pretend to be an out-of-work plumber hoping for a fresh start. I was doing volunteer work to ease into the place and also because I thought it would look good on my CV. The film crew, so the story

TV AND ME

went, was there because they were doing a documentary about volunteer work. Everybody bought it. I guess I shouldn't be surprised, because it ain't hard for me to pass as an out-of-work tradesman from London – people are generally more shocked to find out I might have a few quid!

So I lived in this squalid house on a council estate that was rough as you like. It reminded me of the Rockingham. I got flea bites. Funnily enough, though, I say everybody bought the story, but when you're followed around by a camera crew on a council estate, you get swarmed by gobby little hooligans all shouting 'Secret Millionaire!'. The crew were used to this. They can't actually lie about it so when asked they'd say, 'Oh, if I had a pound for every time someone asked me that', which generally worked. I wasn't so nice. When they asked me, I said, 'If I was a millionaire I wouldn't f***ing come to a place like this, now, would I?' which also worked!

It was arranged for me to work with three different charities in Warrington. One was the Long Lane Garden Centre, which provides work for people with learning difficulties and mental health issues. Another was the John Holt Cancer Support Foundation, which helps those with cancer, and their families. And the third was the Honey Rose Foundation. This one gives people who have terminal illnesses what you might call their dying wish – one thing they'd really love to do before they die, like white-water rafting or a stay in a fancy hotel.

When I heard all this I knew that the producer, whom I'd trusted, was a truly evil man. I was terrified! Sickness, death, frailty of any sort – I avoided these things like... well, like the plague. I'd never – touch wood – been called on to deal with them, and I'd always told myself that if they came too close I had enough money to push them away. I said to the producer, 'I ain't sure I can do this.' Go into some geezer's house who's dying? Would I be sympathetic enough? And adults with learning difficulties? With my sense of humour I'd be sure to say the wrong thing! Couldn't we just go straight to handing over the cheques? That bit's way easier than spending time and working with actual people.

'No,' he said.

So I had to do the thing.

Eight days I spent working with these people. One of my jobs was to drive a woman to an airport for a ride in a helicopter – her dying wish. I thought, 'Man, I could probably afford my own helicopter if I wanted it badly enough, and here she was, saying goodbye to the world with a ride in one.' She was really brave, tears running down her cheeks – I didn't know where to look.

I went to one fellow's house, he'd lost his wife to cancer, and I had to go round and collect her clothes! And I couldn't just bustle in like a plumber, grab the bags and run, either; we sat down on the settee and had a chat. Then he started crying. His daughter was there, crying, too. That was very hard.

People with learning difficulties – totally unknown territory for me. I never thought I'd have the patience to work with them. The garden centre guy, Peter Tickle, told me he loved working there. He said he was sad when the day finished and couldn't wait to get back in the morning. I thought, 'Mate, that's a bit over the top, isn't it?' But it turned out to be the most rewarding of the three experiences. I didn't think I'd be caring enough, but after hanging out with these guys for a while, working with them, eating with them, I started to see what Peter meant: they were really nice people. They know how to enjoy themselves and they're grateful for your input; they made it easy.

Then Peter blew me away by saying that they really liked me. I'd come in and they'd be going, '*Charlie*! He's here! Hey, Charlie!' Peter said that was a big compliment because they didn't take easily to new people because of past experiences – unkindness and having the piss taken out of them. I used to mess up what I was doing on purpose. We were making these wooden figures and I made a big deal about how good I was at woodwork and then let mine fall apart. They thought that was hilarious! I made friends for life there.

So I was enjoying myself when the time came to make the big revelation. They give you a script for it. You have to say, 'Pete, there's something I need to talk to you about. Have you got a couple of minutes?'

He says, 'Sure,' or whatever and you say, 'You know I told you I'm a plumber?'

'Yeah,' he says.

Then you say, 'Well, I *am* a plumber, but I'm a very successful plumber.'

Then Pete should go, 'Really? Oh right, that's nice,' or whatever.

Then you say, 'To be honest, Pete, I'm so successful that I'm actually a millionaire, and I want to give you something.'

And once you say that you have to shut up. You mustn't talk because if you talk, they won't, and it's their reaction the producers want to get – that's the highlight of the show. I had

BOG-STANDARD BUSINESS

to tell three lots of people and they were all in tears. They want you to cry, too, if the mood takes you, but I didn't. I could have, but I kept my emotions in check. Generally, I don't go around crying in front of people, let alone on camera! I did have to walk away a few times, though.

I gave each of the charities twenty grand. The producers prefer it if you give one of them more than the others, for the drama of the choice, but I couldn't do that even though the garden centre was my favourite. They used their cash on a new workshop, a new toilet block, and to provide woodworking courses to the public. Since then I've been up to deliver a BBQ as a Christmas present and I invited them all to come as special guests at one of our Christmas parties. The Honey Rose Foundation used their cash to provide theatre trips, family holidays, and other special things for people with terminal illnesses, while the John Holt Cancer Foundation used theirs to secure new premises because they were being chucked out.

Without a doubt I took loads more away from the whole experience than the charities did. It brought me down to earth and helped me get back to being a proper human being. It made me go, 'Hold on a f***ing minute! Forget your car, forget your nice house, forget your money, this is what it's all about.' One of the top ten things I ever did, it made such an impression on me that when I'm in First Class on an airplane or at some swanky function I catch myself looking around, thinking, 'I bet none of you lot have done *Secret Millionaire*.' I wouldn't mind having it on my gravestone: 'Charlie Mullins, Secret Millionaire'. (What I used to want on my gravestone was, 'What are you looking at?')

People often ask me, 'Charlie, doesn't all this TV stuff get

TV AND ME

in the way of running your business?' My answer is, don't be daft! For one thing, because we're a family business, I have my family's support in running it. Lynda's in the call centre, Scott's operations manager, Samm's a top plumber, Lucy's in accounts and Alice is in the call centre. I'm everywhere! The other thing is, what else is my job, if not increasing our profile?

Publicity creates a virtuous circle: the more publicity we have, the more people hear about us. The more people hear about us, the more customers we get. And the more customers we get, the bigger and better our workforce becomes. The bigger and better our workforce is, the happier our customers are. And on it goes. How else do you think we manage to grow our sales by 15 to 20 per cent a year? If I have to take half a day out to appear for three minutes on a talk show or a news programme — three hundred grand's worth of advertising, by my reckoning — I'd be justified in taking the rest of the week off! What better thing should I be doing?

So when I'm asked at conferences, 'What's your secret for PR?' I say this: Get noticed. Do big things and little things, silly things and serious things. The potential to stand out from the crowd is huge, and it's also available to everybody. You just need to screw on your publicity head, tight, and keep it there.

My publicity head has led me to do some outlandish things, and they tend to work. Probably the most outlandish thing was me, a plumber, hiring the UK's best-known, most expensive — and now utterly disgraced — publicist.

CHAPTER NINETEEN

THE PUBLICITY MACHINE

I was shocked when the news came out in December 2012 that Max Clifford had been arrested for sexual offences. We'd stopped using his company only earlier that year, and everybody was still numbed by the sexual abuse scandal that had broken out over the radio and TV presenter Jimmy Savile. I have to admit, I had real trouble squaring what I was hearing about Clifford in the news with the man I'd worked with for nearly five years. There was nothing about him to suggest that he was anything other than what he appeared: a straight-dealing and highly effective publicist. Then again, I wasn't a vulnerable girl, was I?

For a long time I wrestled over whether to include this part, or not. The last thing I want to do is glorify the man, and my thoughts are only with his victims, whose lives he damaged so badly and who had to go through hell all over again in the courts just to get some form of justice, after all those years. But it was the publicist who turned out to be criminal, not the publicity he was able to generate, and that publicity is an important part of how we got to where we are now. My goal is to show how the mechanics of all that worked, because it was all new to me. So I hope you, the reader, will understand where I'm coming from when I describe it.

The sort of people who can afford Max Clifford's fees — which were around thirty grand a month when I went to him in 2007 — are not usually plumbers. They are usually movie stars, pop stars and celebrities of all types who need to be in the limelight and can afford it. So Max was more than a little bemused to find himself sitting down to talk business with me, a plumber. But as I said to him, I needed to be in the limelight too, and I could afford it, so why not? I also said, 'Max, you're the best at what you do, I'm the best at what I do, let's see what happens.'

I liked his approach. He said to me, 'Charlie, you'll need to be ready at a moment's notice for just about anything.' 'Okay,' I said. He said, 'Charlie, I don't do contracts. We work together for as long as it feels right, and when it doesn't anymore, for either of us, off we go.' 'Fair enough,' I said.

Then we negotiated fees. I think he was curious about being a publicist to a plumber because he cut me a deal, though it was still bundles of money. 'Charlie,' he said, 'I know it's a lot of money but I can guarantee, you're going to get more out of this than I will.'

'Say no more,' I said, and we shook on it.

Overnight, things changed: Max made things happen. Within days I was in *OK!* and *Hello!* magazines. I was on James Whale's

THE PUBLICITY MACHINE

talkSPORT radio show. He'd been right — inside a couple of weeks I'd had far more publicity than the monthly fee would have got me in advertising.

I was also moving in circles unheard of for a plumber. He opened all the doors. I was getting invites to the Royal Variety Performance. And I was turning up to functions that somebody like me would never in a million years get into, meeting celebrities, having my picture taken with them — it was mad!

He'd say, 'Come over to my office, I've got Muhammad Ali here and he wants to meet you.' (I kicked myself on that occasion because I was in Marbella!) Or he'd say, 'Come join me at this *Hell's Kitchen* filming,' and I'd be on TV, having dinner with this or that celebrity. Eventually I got used to meeting people I'd only ever seen on the telly — Des O'Connor, Bobby Davro, Chris Tarrant, Piers Morgan — but it took a bit of time to learn how to just be myself around them.

So much about PR I learned from Max. If somebody came up to him while he was having dinner and asked for a photograph or something he'd smile and say, yes! I'd say, 'Doesn't that bother you?' And he'd say, 'Of course not, that's how word spreads.' He told me, always get a picture, which is why we employ our own photographer to this day. He also taught me how to double up. He'd get me and Stacey Solomon, say, to have a picture together and bang, it's in *Hello! Magazine* — meanwhile, he's getting money from both of us. 'How clever,' I thought. 'I used to do that running errands for the shopkeepers in Camden!'

Everybody he knows is a celebrity, so he managed to get us quite a few more celebrity customers, as well, like Simon Cowell. I reckon our celebrity client list went up by around 20 per cent, thanks to him. He was pushing Pimlico on all levels. He'd say, 'Charlie, get some plumbers round to so-and-so's house, quick', and the next day it would be in *OK!* — you know, 'Ooh, look who's getting a new bathroom in!'

From thirty or forty years in the business he could see the media potential in his clients, and he was always realistic. He'd say, 'Charlie, you're never going to be front page, but I can get you on the inside pages,' and that was fine by me.

We parted company in 2012, before he was arrested. As they can in any business relationship, things had gone stale between us by then. I found more and more it was us hassling him to make something happen — in other words, to earn the pots of money we were paying him. Now Clifford is finished, and his work and everything he stood for has been corrupted, which is ironic for a man who was once so talented at building up a client's image. All his skills, contacts and arrogance proved to be useless when his chickens came home to roost.

Nevertheless, the experience cemented in my mind just how powerful PR can be. We now use PHA Associates, run by Phil Hall, a venerable Fleet Street mover and shaker and former editor of *Hello! Magazine*. Phil is great! We also still use Recognition PR, who consistently get us in front of yet another sort of audience.

I haven't tallied up the amount I've spent on PR over the years, but every penny was worth it. People say, 'Sure, Charlie, it's fine for you to go on about publicity, but not everyone can afford a Max Clifford.' To which I say, 'Hang on, I could afford Max Clifford because I had a multi-million-pound turnover company,' and I had a multi-million-pound turnover company

THE PUBLICITY MACHINE

in large part because of all the publicity I'd bothered to get. You could have the best plumbing company in the world but if nobody knows you exist it isn't much use. PR is like suits – it doesn't matter what your budget is, if you need a suit there's one out there to match it. The question is not, 'How can I afford to do PR?' but 'How can I afford *not* to do PR?'

So that's PR and me. Meanwhile, being the World's Best-Known Plumber does have its downsides. People get jealous, and some want to have a pop at you. Just as very tall guys can attract the wrong sort of attention in a pub on a Saturday night, Pimlico Plumbers began attracting the wrong sort of attention as we emerged as London's biggest independent services company. There's always someone who fancies his chances, but that's fine – I'll step into the ring with anyone.

CHAPTER TWENTY

NOW THAT WE'RE BIG

If I'm being polite I would say that Steve Cosser was a very, very rude and arrogant man. And if I'm being less than polite I'd say he was a tosser.

He was supposedly this business big shot in Australia, where he'd founded a pay television company that failed and then a wireless telecoms company that also failed. By the time he started bothering me he was living in London and had been a customer of ours, apparently.

I'm always polite to customers, but I get a call from this guy out of the blue and right off the bat he's saying, 'Look, me and you need to get together.'

'Why is that then?' I said.

'I want to buy your company,' he told me.

'You f***ing won't!' I said.

And that's how we were introduced – not a great start.

BOG-STANDARD BUSINESS

That was in 2007, the same year he resigned as chairman of the wireless company he'd founded. I found that out later, and also that, by the time he'd left, the company's share price had sunk by around 75 per cent.

Next thing I know he's gone and set up a plumbing company in Pimlico, called Service Corps, and he's done it by poaching around ten of our guys. He even tracked down our former call-centre manager, who breached his contract terms by joining Service Corps. Word was, he was giving them shares in the company and paying them £1,500 a week basically to sit around and do nothing because he hadn't got any customers. His plan was as old as the hills – launch against me and hurt my business enough that I'd be motivated to sell.

He called again.

'This is Steve Cosser,' he said. 'Do you know who I am?'

'I ain't got a clue,' I said.

'Yeah, you do know,' he said.

'Well, who are you then?'

'I've set up a company with a lot of your blokes, and if you don't sell, I'll put you out of business.'

'You f***ing won't!' I said.

Another time he called me and said, 'You made a mistake fighting me. We need to sort this now.'

'I'm in Marbella,' I said.

'I'll fly out to you right now,' he told me.

I said no, I'll be back in London in a few days, I'd call him. When I didn't, he came in person to the office.

'You're taking on the wrong one, mate,' he said.

He was supposed to be worth £700 million or something.

NOW THAT WE'RE BIG

I had him checked out and the best I could peg it at was £35 million, but I couldn't be sure. All the same, I was worried. He'd lay on a weekend in France for my plumbers – wine, food and lovely accommodation – and then present a contract promising the world. And they were signing it! I'd built up my business over twenty years with hard work, hard knocks and a dedication to quality, and here he was, taking it away over a weekend. How do you fight back against bribery?

As if this weren't dirty enough, he used former Pimlico people to get his hands on our customer database, and then started nicking our customers, including celebrities. That was the last straw; enough was enough. Industrial espionage, this was. If somebody wants to go head to head with Pimlico Plumbers, and set up in our patch, and try to offer a better service with lower rates, more power to them! That's competition. It's good for consumers and it keeps us on our toes. If they want to poach my plumbers with higher salaries than I'm prepared to offer, all I can say is, best of luck. That, too, is legitimate competition. But if they start playing dirty and nicking what's ours in contravention of the law, I won't stand for it.

I called in my lawyers, Mishcon de Reya, and we sued them for a million pounds.

As I said at the time, this was not about the money but about principle. Cosser was a bullyboy with a taste for easy cash. In the world of business he'd never built anything solid or lasting, and for some reason, he'd grown jealous of what I had and decided to try and take it off me with dirty tricks. The money, if we won, would go to charity. Mishcon de Reya even said they'd donate a percentage of their fees – quite a thing for lawyers!

BOG-STANDARD BUSINESS

We launched the suit in 2009. I was stressed; I had no idea what kind of a war chest Cosser had to fight this, and meanwhile, he was still siphoning off my talent. A few big-name and long-time customers were going over to his side, too. His gimmick was to run Service Corps along military lines. I'm not sure what that was all about.

As it happened, he lasted only two rounds. Round one, in July 2009, the High Court ordered Service Corps to disclose all the evidence in its possession, including its IT equipment, and to hand it over to be forensically examined.

Round two came in November that year when it emerged that, true to form, Service Corps had dodged and hidden some of this equipment. 'The rat is smelling very bad,' said the judge, adding that Service Corps' conduct amounted to 'a serious contempt of court, which carried a criminal penalty'. The court ordered an independent expert to investigate the concealed evidence, and also ordered Service Corps to pay our legal costs – £24,000 – within twenty-one days.

Each of those twenty-one days came and went, of course, without a penny coming from Service Corps. Now that made the situation interesting. The tables had turned: because the court had ordered them to pay our legal costs, we were their creditors. I knew his business was a sham. He had a few of my customers but nowhere near enough business to pay the great whacking salaries he'd promised the guys he poached from us. I'm afraid the old boxer's killer instinct came back with force. Just before Christmas we filed a petition to have Service Corps declared bankrupt – I wanted to wipe it off the face of company history.

NOW THAT WE'RE BIG

Just after New Year, 2010, a sign appeared in their shop window on Moreton Street in Pimlico. 'Due to Pimlico Plumbers, <u>Charlie Mullins</u>,' it said (underlined just like that), 'we are no longer trading.'

'We will be however,' the sign went on, 'back in court, September 2010 where we are confident we will win the court case and justice will be done!'

But that never happened. The winding-up hearing took place in February, without anyone from Service Corps even showing up. In the end Cosser crumpled like the cardboard tycoon he was. In all, around twenty plumbers jumped my ship for his — they won't be coming back.

We didn't get our million, but Mishcon de Reya kept their word and refunded £5,000 in fees I'd paid them to fight this case. That tasty sum went straight to the lads at Long Lane Garden Centre in Warrington to help pay for the new minibus they needed — so Steve Cosser turned out to be of some use after all!

I don't know where he is now. Last I heard he'd staked a claim on land in Sierra Leone, Africa, for some mega iron-ore mining wheeze, and was in court there, fighting a rival claimant. 'Typical,' I thought. The mega mining scheme never happened.

Meanwhile, there have been some rather more agreeable expressions of interest in Pimlico Plumbers. While the Cosser thing was still rumbling on I was approached by Jim Zockoll, founder of Dyno-Rod.

Now, Jim I liked. A former Pan-Am airline pilot, he'd founded his company in 1963, following a stay in a London hotel that was having drainage problems. He knew about some advanced techniques involving electromechanical processes that were being used in the US, so he got on the phone and arranged to have this equipment sent over right away. It worked. Obviously the hotel was pleased, and so was he, because he'd stumbled on a perfect niche market. Jim and I share a lot of the same values, including the importance of a good strong visual brand – although I think the Dyno-Rod orange vans are a step too far, even for me!

In 2004 he sold Dyno-Rod to British Gas for £57 million, plenty, you'd think, to retire on and forget about work. But money-making is in his blood and, in 2010, at the age of eighty, he rang me up. *The Financial Times* reported in May of that year that I was hoping to raise £25 million by selling a 30 to 40 per cent stake in the business to Jim. The money would be used to expand beyond the M25, and even go national. What the *FT* didn't know was that Jim wanted to do it using a franchise model, which is why the deal fell through in the end.

Franchising is when you offer your trademark, business model and 'special sauce' for sale to licensees. Jim was a franchising fanatic and pioneer in Britain. Before Dyno-Rod the only type of business that operated on a franchise model was fast food, like the Wimpy burger chain. Now you can buy franchises for just about any sort of business. Jim was so keen on franchising that he was a founder member of the British Franchise Association. So naturally, when he came to me, franchising was what he had in mind. He pictured Pimlico Plumbers' franchises sprouting like mushrooms all over the land, netting him and me millions.

But I wasn't so keen. A franchise worked with Dyno-Rod because the Dyno-Rod special sauce was a process for drain cleaning you couldn't get anywhere else. Pimlico Plumbers isn't like that. Plumbers are two-a-penny; our special sauce is a culture and a philosophy perfected over decades. You can franchise the uniform and the liveried vans, but you can't franchise the soul of a business. To me the danger was clear: you'd get any old slapdash plumbers dressed up like us, which would lead to the degradation of the Pimlico brand. It would ruin us. That's why the deal never happened.

The other approach we've had was from none other than British Gas, and this still boggles my mind, and I'll tell you why.

In 2013 British Gas's residential services division — the bit of it that competes with us on boiler repair and maintenance — generated gross revenues of £1.65 billion. Mind you, that's for the whole country. I don't know how much they turn over in London, but you might get a rough idea if you break it down by population. London's 8 million people represent around 13 per cent of the UK's 63 million souls. And 13 per cent of £1.65 billion is just over £215 million. Compare that to our turnover of around £18 million at the time and it's obvious that you have a David and Goliath scenario. If my calculations are anywhere near the mark, British Gas in London are twelve times bigger than us.

And yet, they worry about *us*! In the heating services market, we're their main competitor. How do I know this? Because we're inundated with applications from their engineers. Many don't pass muster, I have to say, but we do cherry-pick the best. In the first month of this year alone we took on four top British Gas guys. What they tell me is that the higher-ups talk about us all the time. What Pimlico Plumbers is doing comes

BOG-STANDARD BUSINESS

up regularly in management meetings: they see us as a threat.

Soon after we moved to Sail Street they came to see us, two of them, the operations manager and their mergers and acquisitions guy. They were very interested to see our call centre and how we did things, and I was happy to show them. They were quite upfront, saying they'd love to offer a premier service like we did, but they were some way off from being able to. Is there a way we could work together, they asked. Maybe your call centre could join in with ours? We could take all your calls? 'No, thanks,' I said. Well, maybe we could send work your way that wants a premier service? 'Thanks, but we get plenty as it is,' I told them. (Why pay them to get work?) How about we amalgamate in some way? (Buy me out, in other words.) 'I'm flattered,' I said, 'but no thanks.' They left empty-handed.

Why are these people drawn to us? What is the secret formula that they've tried to steal, or buy, or persuade us to share with them? I've always been too busy to give it much thought, but doing this book has given me the chance to sit back and reflect. And all I can say is, there is no secret formula. If there is a formula, I guess it would be this:

1. We do the blindingly obvious, and we do it better than anybody else.
2. We make money from happy customers.
3. We require our people to do things our way, exactly, and to have the correct attitude, always.
4. If there is a battle to be fought over the above, we are ready to fight it, and we don't compromise.

NOW THAT WE'RE BIG

5. We don't bother with fancy business theories or business-school gobbledygook: we expect quality work, and we reward results.
6. We try new things and, if they work, we go for it.
7. Our people are motivated, because they are part of a successful company.

So, the good news is, any company can do this, even brand-new, little companies. What gets the wannabes and the get-rich-quick crowd down is that it can take a lifetime of passion and hard work for a good company to grow to any size and for its brand to be untouchable.

You might ask, why not do as Jim Zockoll did: sell up, take the cash and retire? The answer is pride. Pimlico Plumbers is an excellent little company and we haven't even begun to realise our full potential – and that's within London, let alone the rest of the country. Compared to British Gas, our engineers are better and our service is better. They're a cog in a massive corporate wheel and we can outgun them every time when it comes to the most important thing – the customer experience. Why would I give all that up for a pot of cash I don't really need?

There is a rumour on the plumbers' grapevine that British Gas is thinking about setting up some kind of crack squad of heating engineers in London to up the company's game. Rumour has it that I should be worried. But I'm not. I do hope they set up this crack squad – it will be good for their customers and it will give me a new pool to fish for talent in.

Meanwhile, we're just going to keep on doing what we've always done – the blindingly obvious. And we're going to do it better than anyone else. I am very excited, for instance, about our latest venture. In October 2014, after years of me hemming and hawing, we opened our very own plumbing and heating supplies shop, just across the road from our headquarters on Sail Street. Pimlico Plumbing & Heating Merchants, it's called, and I love it!

Now, I can understand that it will be hard for people outside the plumbing fraternity to fully appreciate my excitement about this, so I will try to explain. To serve our customers, our plumbers spend millions every year on plumbing and heating supplies. In doing this, they have to queue up behind members of the public. They must suffer being served by sales assistants who know nothing about plumbing or heating, often to be told that the part they need is not in stock. When it is, they pay through the nose for it.

It wasn't always like this. Proper plumbers' merchants used to be the norm. They had everything, and they were run by people who knew their stuff. They did not serve the general public. Plumbers could go in, get served promptly, and get on with the job. They would also be charged at 'trade rates', which were lower than high-street retail prices.

But when the big DIY chains came along, usually in out-of-town places that were easy to drive to, a lot of the old-style merchants struggled and, over the years, shut down. So now plumbers have to use the same places the public use. If you're a plumber, you feel a special kind of fury when you wait half an hour for a sales assistant to coach some do-it-yourselfer through buying a washer, while you need a three-piece bathroom suite so you can make your living. When I was on the tools you still had

NOW THAT WE'RE BIG

hardware shops with 'trade-only' counters, where you could expect a modicum of priority treatment and some expertise, but even those seem to have disappeared.

Well, we're bringing it all back. Even the name, with the 'merchants' in it, has an intentionally retro feel. It's going to be run by a knowledgeable crew of our own people, led by my son, Samm. It will have everything a plumbing or heating engineer needs, with a proper, high, wide, sturdy counter so the customer can lay the gear out to examine it. (Try finding a proper counter in a DIY branch store these days. All you've got, somewhere up at the front, are tills.)

It will also (eventually), and this is the clincher, be open twenty-four hours a day, seven days a week. The reason for this is that plumbing is a 24/7 service. We cover thirty emergencies every night. Over Saturday and Sunday we'll do 200 jobs, easily. Most of these happen when the shops are shut. What it means is that nine out of ten emergency jobs require the plumber to return in the morning, or on the Monday, when he's had a chance to get the necessary part. He can deal with the emergency – stop the gas or the water flowing – but to get everything working again usually requires something they don't carry. Plumbers carry the basics with them in their vans, but you'd need a train to carry everything. With this shop, our 24/7 service will be completely 24/7. There's nothing like it in London now. This is one of the biggest, most advanced cities in the world, and you can't get a washer on a Saturday night.

I've been thinking about this for years. At one stage I was taken with the idea of a 'drive-thru' plumbers' merchants. But plumbing and heating supplies are not hamburgers. Plumbers

BOG-STANDARD BUSINESS

like to check what they've got, the quality and quantity, because any mistake means a wasted trip, which is no fun at two in the morning.

Originally, I was thinking our new shop would just serve our plumbers, but there are a lot of plumbers out there who would find this service handy, so we opened it up for their business, too. (We're going to be very selective about our customers, though. Anyone who has tried to cheat us, or plumbers who left us owing money, we won't serve.)

It's going to really change things. I know this because of the waves it started making before it was even open, when we were still doing the refurb. Plumbers gossip like anybody else, and once the rumours started trickling through the pipeline, suppliers got the jitters. We got calls from most of them; they were very anxious to supply our shop. I went in to a local branch of a national hardware chain and the sales assistant asked me, was it true about us setting up a plumbing and heating shop? I said yes, it was. 'Well,' she said, 'that's us finished!' We had plumbers coming in, saying it was long overdue, that if we ran the shop like we ran the company, it would be great, and could they set up an account?

Who knows how else we might be able to spin off our brand and our services? We get new proposals every other day, some of them quite interesting. All I know is that if we stick to our core principles, we won't go wrong because, as I've said before, common sense ain't that common!

So that's pretty much where we're up to on the business side of things. Now I'm going to say a few words about what this journey has been like for me, personally.

CHAPTER TWENTY-ONE

NOW I'M RICH AND FAMOUS

The last time I met up with Prince Charles, do you know what we talked about? Plumbing! It was at an intimate gathering at Kensington Palace in May 2014 to thank us patrons of The Prince's Trust. We shook hands and he brought up the plumbing at Buckingham Palace, which is very old and tricky. 'It's always good to know a good plumber,' he said, and we had a laugh.

It just goes to prove the old saying: the more things change, the more they stay the same.

Money and fame do change your life. Mostly, they give you access to places and people you couldn't have access to otherwise. They give you nice things, too. But they don't fundamentally change who you are.

Here is an example of what I mean by access. Recently my wife and I went to a West End show, and we booked into The

BOG-STANDARD BUSINESS

Savoy. When we pulled up in the Bentley the doorman came rushing over, all friendly, like we were there every night, and said, 'Hello, sir! Yes, it's no problem to leave your car right here.' My friend and his wife were behind us in a Mercedes. He wasn't even out of the car before the doorman was on him, saying, 'Oi, you're not leaving that there!' I shouldn't have teased my friend about it, but I did. I spent a good few years as a plumber being told by snooty housewives to go round the back, so I do get a kick out of being welcomed like that at The Savoy.

Once the money began coming in, I started spending time in Marbella, Spain. I love Marbella — the weather is nice, the Spanish people are great, and there are lots of English people about, so I feel at home. It's free and easy, and the social life is incredible. So, about eight years ago, I bought a villa there. What happened was, we were invited to a party on this exclusive little development fifteen minutes out of town, just eight villas designed and built by a landscape architect from Belgium, called Robert. The grounds were beautiful, full of exotic plants, and every house had its own sea view and pool. I thought, 'I want a piece of this,' but none of them were for sale. I kept calling Robert, who lived in one of the villas, but he would just say '*Non!*' and put the phone down. Eventually, one of them became available. The owner, an English bloke, had to return to the UK. Robert invited me round, and I bought it there and then. A year or so later, I bought the one next door. Why do I need two villas? Because I've got four kids and eight grandchildren. It's nice for us all to be there.

Of course, I had a Pimlico team come over and do them up. Nothing against Robert, he's a big-picture guy. The plumbing, meanwhile, was crap – a typical Spanish bodge. The pressures were all wrong, and so was the drainage. It was dated, an eyesore, all done with no consideration for quality or style. Now it's fantastic, of course. We've got state-of-the-art showers, saunas that give you massage and aromatherapy (they look like spaceships and I haven't a clue how to work them), and the baths can fit eight people, each. If you want a bath, it's advisable to have a lifeguard!

My favourite restaurant in Marbella is Villa Tiberio, a famous haunt for celebrities and bigwigs. The other day I called up to book a table for me and some guests, and was told that none were available. 'Can I speak to Sandro,' I said. (Sandro Morelli is the owner. He used to be at The Ritz and the Cavendish hotels in London; I know him.) He came on the line and said, 'Of course, come right over.' We arrived and he personally seated us at a lovely table outside. 'Sorry about that, Charlie,' he said. 'I've had to turn ninety people away tonight!'

After dinner we went to Joys, a great live music bar in Puerto Banus, run by the famous entertainer, Paul Maxwell. We were shown to a nice, stage-side table and Paul, also known as the Piano Man of Marbella, interrupted his song to say to the audience, 'Ah, we're lucky tonight, ladies and gentlemen... we have Rod Stewart with us in the audience!' He meant me – he likes to take the piss out of my hairstyle.

So, yes, money and fame give me access to people and places I could only have dreamt about before. Another place is Dubai, where I usually spend a month every Christmas. Dubai is probably my favourite place in the world. I mean, there is the luxury side of it – the hotels are magnificent, and the service is

BOG-STANDARD BUSINESS

amazing. In the hotel we go to, the Atlantis, you can choose from nineteen restaurants without leaving the building. The weather is hot and the private beach is lovely; the shopping is out of this world. But what I really like about Dubai is its brashness, its boldness. They don't do anything there unless it's the first, the best, or the biggest. They put a ski slope in the desert. Why? Because they can. They do a fireworks display and it wins an international prize. They fancy having the tallest building in the world, so up it goes. People say it's tacky, but I say it's unforgettable. And that's just the superficial stuff. Underneath it all is a vision to be an international trading hub. They don't have much oil to speak of, so they had to do something else. They built the seaports and they built the airports. If you're an American trader and you want to break into Asian markets, you go to Dubai. If you're a Chinese company and you want access to India, you go there too. Now they're building the biggest airport in the world – Al Maktoum International. By 2030, it's going to handle more than 200 million passengers and 12 million tonnes of cargo every year. You think Heathrow's big? Heathrow handles about 73 million passengers every year. It'll be dwarfed by Al Maktoum. That kind of thinking I really admire.

Funnily enough, they know me in Dubai, too. People stop me in the malls and say, 'You're Charlie, right? The plumber?'

Being rich and famous has made me think about friendship. Now, everybody wants to be my friend. I've still got friends from years ago, but my circle of real friends – that is, people I have a history with, who care for me because of who I really am, and vice versa – has grown smaller, while my circle of acquaintances has exploded. Here is another example. The other day I was driving in the Bentley down a street in London, and stopped at some lights, just outside a DIY shop, as it happened. Next thing I knew this guy had run out of the shop and was knocking on my window. Said he owned the shop, and how pleased he was to meet me. Would I take his card? Would I like to see his shop? Could he take me out for dinner? And I was just thinking, 'Why? You don't know me, I don't need a free meal. My business has nothing to do with yours.' People just want a piece of you.

I'm also getting more and more approaches from people with business ideas, or those who want advice, as if I've got some kind of magic potion. I used to have a soft spot for that because business is so close to my heart, but then I realised that there were as many cranks and fanatics with business ideas as there are who are just drawn to famous people. Plus, I could spend twenty-four hours a day mentoring for nothing, and where would that get me? I had a guy the other day fly in from South Africa. He had a plumbing company there and wanted to team up with me somehow. A similar thing happened with someone from Luxembourg. Actually, what this one said was interesting: his family once owned a plumbing business there and wanted him and me to start up a new one. 'Why would I set up a plumbing company in Luxembourg?' I asked. 'It's tiny.' 'I know,' he said, 'but everybody is rich. It has the second highest per capita GDP in the world.' Now I did start to think about that, but in the end it comes down to this: there is still so much to do, right here in London.

People who know me know that I am still very much the

BOG-STANDARD BUSINESS

same Charlie Mullins I was before I became rich and famous. All the other stuff is nice, but do you know what? When I go to Villa Tiberio what I really like, what I almost always order, is not some carpaccio of this or velouté of that: it's steak and chips. And where do I feel most at home? Is it hobnobbing with royalty and celebrities? No, it's in the Pimlico Plumbers' canteen, eating a bacon butty and talking shop. What do I get most excited about? Is it thoroughbred racing horses and super-yachts? No, it's my business, same as ever. Prince Charles wanted to talk plumbing with me. He is a courteous man, and probably reckoned that plumbing was a topic that would put me at ease: he was right.

But having said all that, there is one other thing money and fame have brought, and this I take very, very seriously. What I'm talking about here is influence. I'm not saying my influence is huge or anything, I have no delusions of grandeur but some of the most powerful politicians in the land do seem to court my favour, and to care what I think. This provides me with a very rare opportunity, an opportunity of the sort that doesn't come to many people, and in the next chapter I will set out how I have been trying to use it.

CHAPTER TWENTY-TWO

PLUMBING TO POLITICS

I've done a lot of things for publicity, but banging on about apprenticeships is not one of them. As a publicity-generator, the apprenticeships issue is rubbish. That's because it's downbeat and boring. Everybody knows there's a problem, everybody agrees we need to do something about it, but actually doing it is one of those complicated, bureaucratic quagmires. Plus, politicians and business people pay so much lip service to apprenticeships that we're sick of it. But I keep banging on because I believe in it and, when the Tory-led Coalition took power in May 2010, I finally got the platform I needed to make a big noise.

It was interesting how it played out. In December 2009, when the Labour Government was on its last legs, I put out a statement saying I'd seriously consider leaving the country for Spain, and taking my business with me, if Labour pushed through with their plans for a 50p top tax rate. The country was in serious trouble thanks to the financial crisis – which was caused by bankers – and the idea of taxing entrepreneurs and wealth creators like me to pay for the sins of those parasites and crooks in suits made me absolutely furious. It also reminded me of the big-state policies that wreaked so much havoc in the 1970s. I wanted nothing to do with a country like that.

A few weeks later, I had a visit from the Rt Hon David Willetts MP, who was the Conservatives' shadow minister for universities and skills. It was a great meeting, and he really liked my idea for a fully funded apprenticeship scheme. We were definitely drinking from the same teapot on this issue and I felt I could actually work with the Conservatives.

After the election I went to the Conservative Party Conference in October 2010, where I met George Osborne, who promised to take a fresh look at the apprenticeship scheme. Then, in September 2011, I met with David Cameron and George Osborne at a Downing Street business leaders' reception. There were plenty of other people there but I lost no opportunity to get my point across on apprentices.

I think it worked because a few months later I was invited to a meeting with none other than Professor Alison Wolf, author of 'The Wolf Report', a review of vocational education published earlier in the year. She was great. Her report made no bones about how so many vocational courses are useless for young people and employers. Also at this meeting was Gila Sacks, head of the apprenticeships unit at the Department for Business, Innovation and Skills (BIS). This was pretty exciting because these women are the real policy brains – the scripts that the politicians read are written by people like them. We talked for a couple of hours and I got the feeling that it was a new experience for them, talking to someone like me, because they kept asking, 'What do you think?', or 'Why do you think that is?', and even, 'What if we did this?'

I've sat down with Doug Richard, the IT entrepreneur and *Dragons' Den* guy who the government commissioned to write a big review on apprenticeships in 2012. In 2014, I was asked for my tuppence worth on a cross-party commission on apprenticeships in construction, headed up by Tory MP Robert Halfon and Labour Peer Lord Maurice Glasman. I've met up with Kenneth Baker (Lord Baker of Dorking, to you), who was education secretary under Margaret Thatcher, and who is now a big policy thinker on education. I've become what they call a 'pundit'. I speak at conferences and sit on round tables and get interviewed on TV and radio about apprenticeships. And I keep talking to politicians. I host dinners at the House of Commons, and I'm a regular invitee to the Conservative Party conferences and to the Business Leaders receptions at Number 10. More than that, over the last few years pretty much every senior Conservative politician I'd ever heard of – and some I hadn't – have been here, to Pimlico Plumbers HQ, to see what we do and to make speeches. We've had Education Secretary Michael Gove, Skills Minister Matthew Hancock, Chancellor of the Exchequer George Osborne, Home Secretary Theresa May, Lord Chancellor Chris Grayling, Work and Pensions Secretary Iain Duncan Smith, and, twice, London Mayor Boris Johnson – all of them, of course, with reporters and television crews in tow.

BOG-STANDARD BUSINESS

Now, I'm not totally naïve. I know why they want to be seen with me: I'm exactly the sort of businessman Margaret Thatcher stood for. And I'm proud of that. But on this issue there's only so much posing for the cameras I'm prepared to do. I don't want to be just a pundit, or a poster boy. And, in my opinion, there have been more than enough reports and reviews and commissions, which, from what I can see, usually just kick the issue into the long grass and allow people to get away with arguing over details.

I want real change, now.

According to government stats in the fourth quarter of 2013 there were 1.04 million sixteen to twenty-four-year-old NEETs in the UK – that's young people Not in Education, Employment or Training. Cast-offs, in other words. As the years roll by that number – let's call it a million – may go up or down by a few thousand but, unless we fix our badly broken system, I believe the general trend will be upwards.

There are all kinds of historical reasons why we now have a million young cast-offs in our society. Over the years, British companies stopped taking responsibility for training up their future workforces. That responsibility was shoved onto the government, who let the vocational education system get weak and disconnected from industry and business. Our culture changed for the worse, too. People started to think that learning a skilled trade was a second-class way to go. Schools changed, so that students were no longer streamed into separate pathways leading to university on one hand and vocational training on the other. This led to the ultimate madness of Tony Blair saying that 50 per cent of all kids leaving school should go to university.

PLUMBING TO POLITICS

Fifty per cent! To university! What did he think they'd be doing?

I've never met anybody who didn't think that was looney, but just how looney is becoming painfully clear now. Six months after graduation nearly 40 per cent of graduates are still looking for work these days, according to research by Totaljobs.com (February 2014). And the Office of National Statistics (ONS) tells us (November 2013) that of those who do find work after graduation, 47 per cent are doing jobs that don't require a degree – receptionists, sales clerks, unskilled labour, that sort of thing. Totaljobs.com found that 44 per cent wished they'd had more practical training. I bet they did!

The tragedy is, companies big and small all over the land who make things and fix things, like us, can't find enough skilled, capable young people. Everybody I talk to in business moans about how kids today don't have any experience or qualifications but we're still pumping out these beautiful, useless, university-educated youngsters. The ONS says that the percentage of graduates in the population has rocketed from 17 per cent in 1992 to 38 per cent in 2012. So you have people with expensive degrees in things like golf course management flipping burgers part-time – if they're lucky. They're hoovering up all the low-skilled jobs that people who don't get on in school used to stand a chance of getting. Meanwhile, the government admitted in March 2014 that around 45 per cent of graduates would never earn enough to repay their whacking great student loans – meaning we, as taxpayers, will probably end up bankrolling the whole stupid mess!

Something is very wrong at the heart of our society; we're

coming apart. As I said at the beginning, it's like there are two ships, one filled with British youngsters and the other with British employers, and they're sailing off in opposite directions.

Germany has this sorted, compared to us, anyway. Everybody looks at Germany and thinks, why can't we be like that? In Germany they stream kids into academic and vocational pathways, and they start thinking about it when the kids are eleven or twelve. By the age of fifteen around a third of them are heading towards a more academic path and two-thirds are heading for a more vocational path. The vocational path has classroom study and a paid apprenticeship for every pupil.

Everybody works together on raising work-ready young people in Germany. What gets taught in the vocational schools is set and agreed on by employers and unions and the schools. Also, the costs are shared between industry, the government and the trainees themselves. And do you want to know something else that should make us hang our heads in shame? They laid this system down in 1969, and have been refining it ever since.

Is it any wonder that Germany's youth unemployment rate stands at around 7.7 per cent at the time of writing while here in the UK it's around 20 per cent? And I'll stick my neck out and say it's also no surprise that Germany's economy is driven by innovative manufacturing and exports, while ours is driven by low-skilled service industries, household debt and the parasites in the City of London.

So what do we do? Well, none of us would have started from here if we'd had a choice, but here is where we are. We desperately need to wake up and borrow certain basic aspects of the German system, but that's a long-term project.

PLUMBING TO POLITICS

Meanwhile, I believe there is something we can do right now to start steering the two ships back together.

First, as I said to David Cameron, blow up the Jobcentres, because they're useless.

Next, stop giving kids Jobseeker's Allowance. Instead, give the money to companies so they can take on young people as apprentices. And I mean proper, old-fashioned apprenticeships, lasting three to four years, so that when the young people are finished they are productive and capable.

Speaking as an employer, it's a no-brainer. Jobseeker's Allowance is currently about three grand a year, based on £56.80 a week. If you add this to the current £1,500 a year grant for first new apprentices, you're getting close to half of the wage an eighteen-year-old is entitled to – £5.13 an hour. The money could pay the full salary of apprentices for the first year, to get them started. In the second year, the company and the government could split the salary fifty-fifty. And in the final year the company could pay the full salary because by then the apprentice would be pulling his/her weight.

This money would offset the cost of taking on an apprentice by half, or more in some cases depending on the age of the young person. It would make the business case for taking them on very strong. I have twenty-eight apprentices right now and if the government supported me in the way I'm suggesting, I'd take on twenty more tomorrow.

The government spent nearly £5 billion on Jobseeker's Allowance in 2012. When you consider the million NEETs and the 20 per cent youth unemployment rate, it seems a lot of this money is being poured down the drain. The politicians hope that this cash will be a first step on the path to work, but I fear that in too many cases it's nothing more than a first step on the path to benefits dependency. From this perspective it makes sense from the Treasury's point of view, too. Rather than pouring those billions down the drain, it would be a massive investment in the medium term to have small businesses employing and training thousands of new apprentices who are genuinely on their way to being skilled, capable, sorted, tax-paying adults.

I believe this approach would also generate a new wave of entrepreneurship that our country so badly needs, because once you get people off benefits, off the streets, doing skilled work well, with cash in their pockets and confidence in their hearts, a proportion of them will start doing what people have always done. They'll look around and say, hang on, I think I can make a go of this myself — and start their own business.

The cash is the carrot, but the government could also use a stick. Two sticks, actually. One, for the young people, could be this: you don't get Jobseeker's Allowance if you're not doing an apprenticeship. The other, for the employers, could be this: you'll face tax penalties if you don't pull your weight and take on apprentices.

I've been banging on about this for five years now and when I tell policy people and politicians I can see their eyes light up, and then go dim again. Maybe my plan is too radical. Maybe they feel it would upset too many apple carts. Maybe they can hear howls of protest from everybody who has a stake in the broken, unfit system we have now, from benefits defenders to the further education colleges, from universities to British businesses themselves — and would prefer half-measures that sound good but don't do anything. But I believe we need radical action. I admit I'm no policy expert – I'm a plumber, and if I see something's broken, I fix it.

The rioting and looting we saw in British cities in the summer of 2011 was the clearest sign yet that we're heading for trouble. Not everybody raising hell during those weeks were NEETs, but our million cast-offs represent a growing segment of our society that has no stake in the future. They're going downhill, so why shouldn't they take the society that rejected them along for the ride? I know how they feel because I came within a cat's whisker of being a cast-off myself. If it hadn't been for the kindness of Bill Ellis, first, and my apprenticeship, second, who knows where I'd have ended up?

Apprenticeships used to be a lifeline to the future. That was true both for companies, who didn't even question the need to invest in their future skills base, and for those who aspired to be skilled, independent, contributing members of society. Apprenticeships could be that lifeline again.

People talk about 'broken Britain'. I believe apprenticeships could play a massive role in putting Britain back together again. It's one of the big pieces of the jigsaw, and all the pieces go together. To show you what I mean, we have a serious housing shortage. Just to keep up with rising demand, we need to build something like 150,000 new homes each year, more than double what gets built now. How can we possibly build all these homes? In 2013 the number of young people completing an apprenticeship in England fell to just 7,280, half the number in 2009. At the same time, the Construction Industry Training Board is saying that the industry will need more than 182,000 workers in the next three years. Without a ready supply of skilled workers, the industry will do what it has always done, which is take on migrant workers, stoking resentment over immigration. Meanwhile, because we don't have enough homes, house prices and rents will go through the roof, especially in places like London and the south-east, making people homeless.

Is it just me, or can anyone else see how it's all linked? What you get from this lack of joined-up thinking is the absurd situation where, in the first nine months of 2012, Westminster Council spent more than £2 million putting homeless people up in hotels!

So there you have at least five pieces of the puzzle that even someone like me, who dropped out of school at fifteen, can plainly see are connected: housing, skills, immigration, our NEETs, and our huge benefits bill. What would I do if, God forbid, I was in charge? Well, let's see. For starters, I would commit to a major, long-term, publicly-funded home-building programme, stretching far enough into the future so that the industry feels confident about taking on apprentices by the truckload to meet the demand.

Make those apprenticeships fully-funded, to the age of nineteen and beyond, if necessary, and within reason. Give the companies a real say in designing the training, and stop central government trying to micromanage it from Whitehall. It cannot be beyond the wit of man to design apprenticeships that work properly, for the young people and the companies.

Keep in mind these are not just apprenticeships in what they call the 'biblical trades' (because they've been around since the dawn of time) – bricklayers, joiners, plumbers and plasterers. Putting up modern houses and apartment blocks takes an army of skilled, semi-skilled and professional people, including labourers, CAD technicians, architects, designers, surveyors, project managers, engineers of all sorts, installers of all sorts, marketing people, accounting people, IT people, and more. We would also need to make sure that these new homes are energy efficient to the highest standard. They should have state-of-the-art insulation systems and even localised power generation – and there you have a whole new high-tech product manufacturing and engineering industry of its own. Apprenticeships could be a route into every one of these new jobs, even the professional ones.

Now hang on, Charlie, I can hear you say. Wasn't your heroine, Margaret Thatcher, the one who pulled the government out of social housing in the first place? Right-to-Buy and all that? And to that I would hold up my hands and say, yes, she was. It seemed like a good idea at the time, and there were many thousands of people who benefitted from Right-to-Buy. But I also believe that if she were here, looking at the situation now, she would agree with me. It didn't quite work out how she planned, and now we have to do something to fix it. I'm no economist, but I am a businessman, and this plan, from a business point of view, is a no-brainer. The government invests in a valuable asset, which it can hang onto forever, or sell to investors. And it doesn't all have to be social housing. The government could kick-start the private build-to-rent sector, which is tiny in this country compared to other countries, like Germany. Meanwhile, it would make other good things happen, as well. It would make skilled jobs available, and train up British young people to take them. The benefits bill would come down, homelessness would drop, and the housing crisis would start to ease.

And this is just one sector, housing, which I'm kind of familiar with. What about manufacturing, IT, healthcare, oil and gas, financial services, retail, catering and hospitality? The list goes on.

To me, apprenticeships are not just about getting people off benefits. It's not some short-term, isolated, or cosmetic issue: it's about creating a strong society and preparing for our future. It's connected deeply to just about every other issue we worry about.

So now I say to David Cameron, David, I have vast respect for what you've done and for the Conservatives – I am Tory through and through – but I do believe you are ducking the issue. The chats and the visits with politicians have been nice, but the time for action has come. My great hope is that you will implement a fully-funded apprenticeship scheme without delay.

In putting together this book I was asked, Charlie, what if the Labour Party adopted your plan, and the Conservatives didn't? Would you switch allegiance? I was dumbfounded. What a thought! But after a few seconds I said yes, yes, I would. I would have to, because I would owe it to our future.

Let's hope it never comes to that.

There, I've said my piece. It has been a roller-coaster ride getting to this point – a lot of fun and a lot of hard work. But I'm going to give the last word to my mentor, Bill Ellis. I never forgot what he said, and I hope you will pass it on to whoever will listen: 'Charlie, if you learn a trade you will make lots of money and you will never be out of work.'

THE LETTER FROM THE QUEEN THAT I ALMOST DIDN'T OPEN

I thought I'd finished doing chapters for *Bog-Standard Business* a few months ago, and all I now needed to do was sit back and wait for the books to come in from the printers. But here I am on 5 January 2015 writing another chapter, telling you about the OBE I have just been awarded by Her Majesty, for services to plumbing. Let me say right up front what an honour it is, and one that I never in wildest dreams would have believed could happen to me.

Who would have imagined the Camden street urchin, born in the same year as the Queen came to the throne in 1952, would one day be invited to Buckingham Palace to be made an Officer of the Most Excellent Order of the British Empire? Not me that's for sure.

And it almost didn't happen, since at first I didn't want to open the letter bringing me the good news. The thing is, in my

BOG-STANDARD BUSINESS

experience letters that arrive stamped HM Service usually only mean one thing, and it ain't good! Generally it's the tax man and he wants money off me!

It was my daughter, Alice, who talked me into opening it in the end. And I can honestly say it's the first time in my life I've been genuinely lost for words. I think she thought it was a really huge tax bill by the look on my face.

Surely it couldn't be true? They didn't want money, they wanted to give me a medal that only rich and famous people get. You know, Olympians, pop stars, writers, even . . . not plumbers from south London. But it was real, and even after reading the letter three times over I couldn't find the catch, so it must've been true.

Having had a few days now to reflect on the latest 'chapter' of my amazing rags-to-riches life story, I think the nice thing about this great honour is that they've given it to me for 'services to plumbing'. I can't imagine many people have ever received an OBE for plumbing.

I am however very proud to be perhaps the first plumber to be honoured by a reigning monarch since Queen Victoria presented Thomas Crapper, another former London plumbing apprentice, with his first royal warrant.

And just like perhaps the most famous of plumbing entrepreneurs the world has ever known, I am incredibly grateful to the profession that has given me so much. This is why throughout this book I have gone to great lengths to stress the importance of training the next generation of plumbers, through real old-fashioned apprenticeships.

The good Mr Crapper was, I've read, also a big fan of the

THE LETTER FROM THE QUEEN THAT I ALMOST DIDN'T OPEN

apprenticeship for training the very best of engineers to continue his company's work. And being awarded an OBE for continuing this proud tradition is perhaps the greatest honour a plumber like me could every receive.

ACKNOWLEDGEMENTS

I would like to dedicate this book to the man who is responsible for getting me into the career that has given me so much, the great and inspirational Bill Ellis, plumber. When your mind is young it's open to all possibilities, and I am very lucky that I ran into Bill when I was just a lad. I'm also lucky he wasn't a bank robber, since this charismatic man was my hero and I would have followed him anywhere.

As a kid my other great love was boxing and if my boxing career hadn't been obliterated in the ring one night, fighting for London against Wales in 1973, this could have been a very different book. Despite my premature departure from that field I still have a lot of life lessons to thank fighter and Fisher Club boxing coach Micky Kingwell for.

A book about Pimlico Plumbers wouldn't be complete without mention of John Harper, the man who first got me on the serious road to building a big and successful company. As owner of Pimlico Properties in the 1970s John leased me a basement office, my first business premises.

I would also like to pay tribute to Tony Davison, my marketing director, who stuck with me through the recession of the early 1990s, when the banks tried to take me down, and everyone else thought I was finished. Tony helped me put things back together when they were broken, and many times he's been there for me as a sounding pole when I have faced difficult decisions.

Graham Robb, who so kindly wrote the foreword to this biography, has, with his PR company Recognition PR, been with me for more than a decade, offering sound media advice. Graham has become a trusted confident and friend, and is another person I have turned to for help in tough times.

I would also like to thank Phil Hall, owner of media agency PHA, who finally, after many years of wondering whether I should write this book, guided me through the initial rounds of getting a proper publishing deal, and convinced me to get on with it.

And there's Annie Williams, the former head of corporate partnerships at The Prince's Trust, who brought me into a long-standing association with that worthy institution, which does such great work with kids, many of whom come from a similar back ground as myself.

And of course where would I be without my wife Lynn and children Scott, Samm, Lucy and Alice? They have stood by me for three decades, and helped me build and run the most successful independent plumbing company in the UK.